FIRST EDITION

UNDERSTANDING THUNDERSTORM GENERATORS

MALCOLM BENDALL'S OPEN SOURCED CARBON TO OXYGEN CONVERTER

Compiled, written and illustrated by
AM&ER Publishing Company

AM & ER
PUBLISHING COMPANY

REVISION HISTORY
JANUARY 2025 - "Understanding Thunderstorm Generators: Malcolm Bendall's Open-Sourced Carbon To Oxygen Converter" first released.

ISBN 979-8-218-55250-3

Book Cover and Illustrations by AM&ER Publishing Company, LLC

First Edition 2025

www.AMandERPubCo.com

Dedicated to those condemned for questioning.

"It is the mark of an educated mind to be able to entertain a thought without accepting it."
- Aristotle

"All great truths begin as blasphemies."
- George Bernard Shaw

"Great spirits have always encountered violent opposition from mediocre minds."
- Albert Einstein

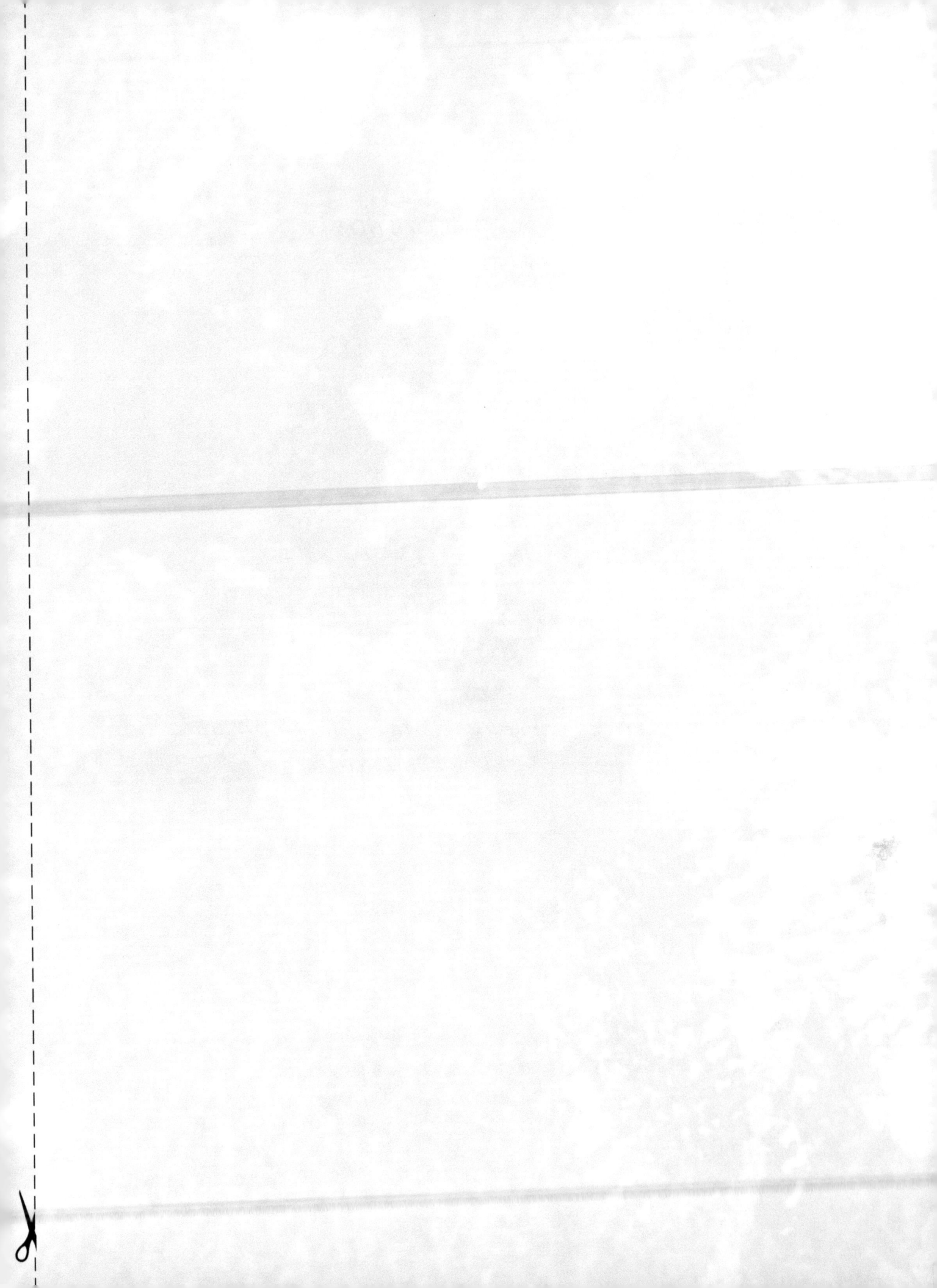

TABLE of CONTENTS

*Explanations for the words highlighted like this.

THANK YOU!

MALCOLM BENDALL

Creator of the Thunderstorm Generator, this Tasmania-born geochemist has a rich indigenous heritage and profound spiritual perspective. He has emerged as a pivotal figure in the global quest for sustainable energy solutions. His groundbreaking invention, a retrofit to conventional internal combustion engines that uses plasmoid technology to passively convert carbon emissions into oxygen, drastically reduces air pollution caused by fossil fuels.

BOB GREENYER

A clean energy researcher and founding leader of the Martin Fleischmann Memorial Project (a group dedicated to the research, development, and global understanding of Low Energy Nuclear Reactions), Bob has independently tested and proven Malcolm's theories. An expert in plasmoid technology, Bob believes everything in the universe follows a natural system of order that can be observed on a galactic scale as much as a molecular level.

RANDALL CARLSON

With four decades of study into the intersection between geology, astronomy, sacred geometry, and ancient civilizations in his toolbelt, this prolific geo-mythologist and documentarian is not only contributing to modern society's understanding of the past, but he also serves as an active crusader in exploring the profound connections between humanity and the natural world.

JORDAN COLLIN

A student of Malcolm's and creator/host of the YouTube channel "Alchemical Science," this mathematician, self-taught farmer, electrical engineer, and physicist presents in-depth videos that break down how the technology works and tracks its ongoing progress.

SHOW SUPPORT

MALCOLM BENDALL'S
STRIKE FOUNDATION
www.strikefoundation.earth

BOB GREENYER'S
MARTIN FLEISCHMANN MEMORIAL PROJECT
www.quantumheat.org

RANDALL CARLSON'S
PERSONAL WEBSITE
www.randallcarlson.com

JORDAN COLLIN'S
ALCHEMICAL SCIENCE
www.alchemicalscience.com

ABOUT

↳ AM&ER PUBLISHING COMPANY

AMandERPubCo.com

IN EARLY 2023, a transient culture writer/editor and a traveling sound mixer/illustrator met for the first time at a hipster sushi bar in downtown Phoenix (thanks, Hinge). Over the next five hours, the duo bonded over short rib bao buns and a shared affinity for absurdist comedy and original storytelling. It's such sweet serendipity when lifestyles, humor, growth journeys, and creative interests align! Amanda and Eric supported and fueled each other's art naturally, and they would often stay up riffing, rambling, and spitballing story ideas until 3 a.m. One day, they decided to turn their favorite musings into shareable stories, launching AM&ER Publishing Company in the spring of 2024.

AM&ER PubCo. is a chest full of curios: offbeat stories, fringe science deep-dives, dark children's fables, freestyle poetry, steamy romance novels (maybe?), naughty adult coloring books (let's try it), and an insomnia-induced short story about the existential journey of an armless robot who was born to hug (one of our best). They're each little slices of life that we hope you enjoy as much as we did dreaming them.

~ AM

(Illustrated by ER)

09

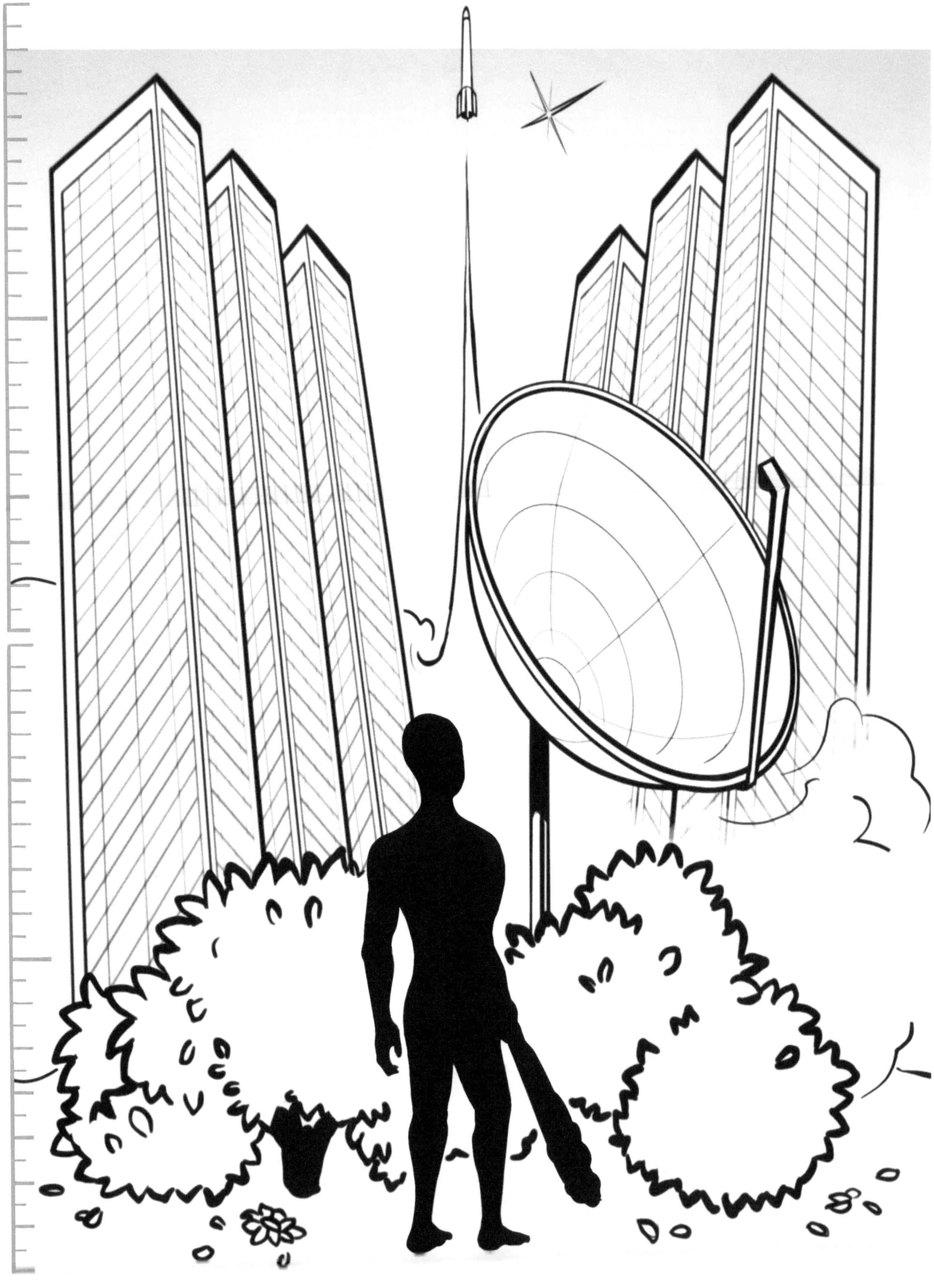

FORWARD

Have You Ever Watched A Sci-Fi Movie and pondered what it might be like to live in that society? Well, Malcolm Bendall's Thunderstorm Generator might be your ticket to the future. It's a revolutionary technology with the potential to change the world. By recreating the dynamics of natural thunderstorms, the system targets the toxic gases generated by conventional combustion engines (like the one under the hood of your car) and turns them into breathable air, using - and this is key - barely any processing power. Imagine if the emissions from your tailpipe consisted of little more than pure, breathable oxygen. It might sound like science-fiction wizardry, but it's very, very real. And the secret lies in tiny, doughnut-shaped balls of energy called plasmoids.

What's The Big Deal With Plasmoids?

If you've ever taken a physics class, you might remember plasma as the fourth state of matter (in addition to solids, liquids, and gases). Plasmoids are created when plasma is introduced to a magnetic field, producing coherent, self-contained structures that can hold and release a <u>ton</u> of energy. (If you took two pieces of wool and rubbed them together in a dark room, you'd make tiny plasmoid flashes of your own!) Plasmoids are special because they've been shown to convert elements into other elements in a fascinating process called transmutation, and in the case of the Thunderstorm Generator, they gallantly turn carbon gases into oxygen. This concept, while groundbreaking, is far from new.

A Quick History Lesson

Flash back to the year 1900, and Nikola Tesla is widely considered the globe's leading electrical engineer. His development of AC current revolutionized the electrical industry by enabling efficient long-distance power transmission, which paved the way for the modern power grid and wireless communications systems. (That cell phone in your pocket? Yeah, he laid the groundwork.) But Nikola had a much grander vision: to provide a renewable, wireless, limitless source of power to the world. How? By harnessing and converting energy from the environment into usable electrical power in a concept called Radiant Energy. His plans were thwarted, and Nikola's dream of free electricity for all was mostly lost in history.

Nikola wouldn't be the last to think beyond the boundaries of traditional scientific understanding. Wilhelm Reich's Orgone Energy, Ken Shoulder's Charge Clusters, George Messier's Ektons, and Paul Dirac's Monopoles all explore alternative forms of energy and matter. The central theme is that everything in the natural world works in mathematical harmony, and when harnessed precisely and correctly, small pops of intense electrical energy have the power to transfer the building blocks of atoms (protons and electrons) to other atoms, effectively altering their elemental makeup. Malcolm Bendall is the latest in a long line of pioneering thinkers to apply this knowledge in a profoundly impactful way.

How Does The Thunderstorm Generator Work?

Like its name suggests, the Thunderstorm Generator works by creating static electricity, similar to how thunderstorms produce lightning when warm and cold air intersect. It has three main components: Ionization Chamber, Bubbler, and Vajra (more on each later), that ultimately create a meeting point for the hot carbon gases from the car's exhaust and cold, ionized plasmoids, simulating a thunderstorm. When carbon passes through the "zero point," a state of total equilibrium where time is suspended, it re-emerges in the atomic structure of oxygen (8 protons, 8 neutrons, and 8 electrons). Even more impressive, the process demands an extremely low energy input. This is possible by following the principles of sacred geometry, an ancient science that explains the energy patterns that create and unify every single thing in the universe.

What's Next?

The device not only brings a combustion engine's carbon output down to nearly zero, it also increases a vehicle's fuel efficiency by 15-40%. Pretty incredible, huh? And Malcolm doesn't plan to stop at cars. He's thinking big - like nuclear power plants big. His goal is to lead the world into a new era of clean energy, where fossil fuels no longer stain the air with dirty greenhouse gases. The following pages will provide a detailed analysis of how the technology works, along with a DIY guide for building your very own Thunderstorm Generator at home. Let's take a ride into the future, shall we?

A NATURAL SCIENCE IS BEING REDISCOVERED

AS ABOVE

SO BELOW

INSTRUMENT OVERVIEW

The suction throughout the system is provided by the engine.

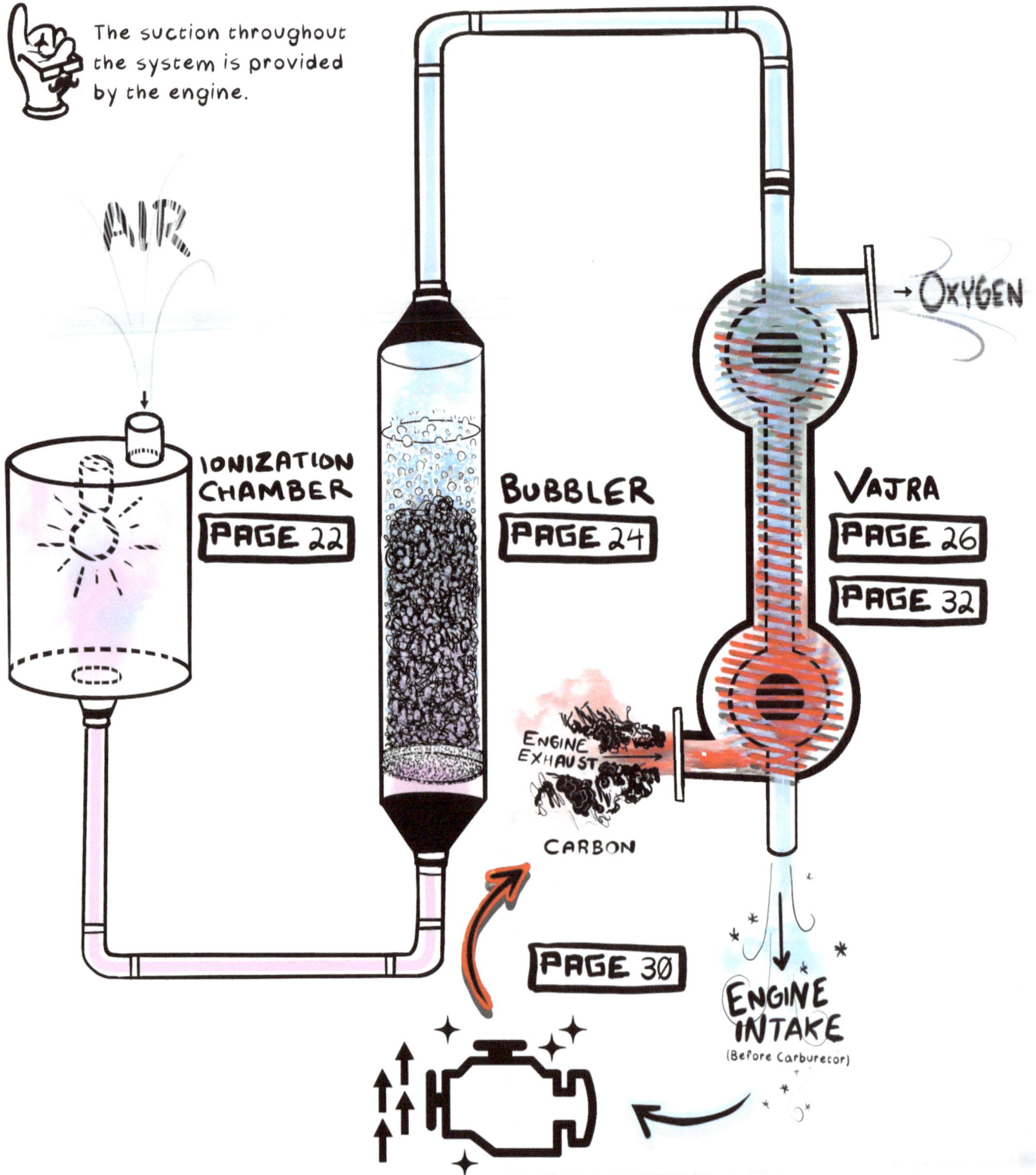

AIR

IONIZATION CHAMBER

PAGE 22

BUBBLER

PAGE 24

ENGINE EXHAUST

CARBON

VAJRA

PAGE 26

PAGE 32

→ OXYGEN

PAGE 30

ENGINE INTAKE
(Before Carburetor)

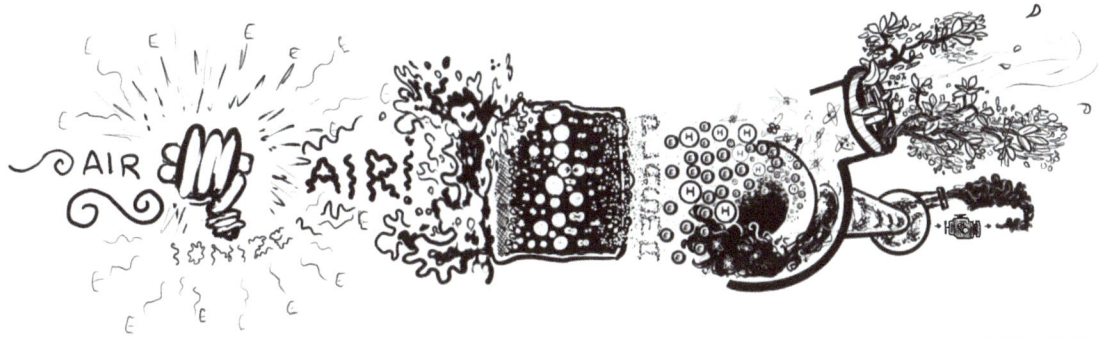

? WHAT'S GOING ON HERE

OK, we're basically hijacking the suction created by the pistons of an engine to treat air through three stages: light, bubbles, and alchemy. Air is brought first into a chamber with an ultraviolet (UV) light that knocks away electrons to create a field of charged particles. It's then pushed through a porous stone (like the ones found in fish aquariums) to create tiny ionized bubbles that implode to form self-contained plasmoids. The plasmoids are swirled into a chamber of precisely measured spheres to convert the exhaust gases into oxygen. We're turning cars into trees using the mechanics of a thunderstorm.

There are no moving parts in this device, and everything you need is cheap and easily accessible. The entire unit utilizes the suction created by the engine's normal operating system, so it can be retrofitted to every. single. one.

The following pages will break down each of the three main components: the Ionization Chamber, the Bubbler, and the Vajra, plus some notes about the engine, to illustrate the mechanisms taking place that passively transmute carbon into oxygen. We'll include explanations, an atomic breakdown, and DIY build instructions for the enthusiasts and home scientists among us.

BEFORE YO

THERE CAN BE NO AIR LEAKS IN THE SYSTEM

The efficiency of this system is dependent on its ability to harness the suction created by the engine's _pistons_. Air leaks in the vacuum will compromise the system's benefits.

+100 NEED ACCESS TO SKILLED WELDING

Crafting the Vajra is the most difficult part of this project because of the importance of the build's geometry and the aforementioned vacuum. You'll need metal parts to facilitate the transfer of hot engine exhaust, but more importantly you need to keep exact measurements to maintain the resonant _frequencies_ produced by the spheres or the system will not work.

OBLIGATORY: BE SAFE

The system is safe, but welding metal and handling untreated carbon monoxide exhaust that can reach temperatures of 392°F (200°C) can cause injury or even death (as stated on any engine's standard warning label). This book's intent is to inform, not harm.

OU START!

THIS SCIENCE BOOK
USES THE
IMPERIAL MEASUREMENT SYSTEM

INCHES?!

Yes! 😊 Inches, feet, and the like. We'll also be using micrometers (microns), which is one-millionth of a meter, so don't worry!

But why, you ask, are we using the imperial system instead of the ever-so-efficient, tried-and-true metric system?

When you look at the values of inches and miles, those same numbers appear in other measurements found in nature and match the harmonic frequencies of matter at an atomic level. Meters don't represent that.

The imperial measurement system is integral to Malcolm's base mathematical theory, which mirrors ancient vortex-based/Sanskrit mathematics. [12, 13] This understanding of numbers is *very* different from what we were taught in schools. For instance, the numbers 8, 6, and 4 appear in the diameter of the sun (864,000 miles), the moon squared (8,640 miles), and the number of seconds in a day (86,400 secs). The first law of Sanskrit mathematics is that a number primarily identifies the dimensions it represents. Zeros and decimal points imply an increased or decreased order of magnitude and a shift in the musical key. An octave below 8-6-4 is 4-3-2, the three diameters used in the spheres of the Vajra that make the magic happen. The Vajra's design is akin to a musical instrument, emphasizing precise measurements that capture the harmonics of a musical scale in the note of A at 432 hz. See? *VERY* different.

SUPPLIES

Here you'll find a full list of all the supplies needed in the following sections.

When selecting which materials to use, stay consistent in your choices (for example, don't mix metals in your pipes. Keep them all tungsten, copper, or titanium). For the spheres, always use 304 alloy steel, for strength and ease of welding properties.

The consistency of materials seems to have an effect on the resonance of the entire device. You want every piece to work in harmony. "Like an instrument, a flute made of silver resonates better than one made of steel." - Jordan Collin

CONNECTING BITS
04x Tubes of sealant (to ensure connections are air tight)
01x Tape, strong enough to securely hold two 4" hemispheres together
05x 0.5" Right-angle air conditioner fittings
07x 0.495" ID, 0.625" OD, 0.065" Tungsten pipes (you can also use
  copper or titanium)

IONIZATION CHAMBER
01x E27 UVC sanitizer ozone-free, 25W 254nm wavelength light bulb
01x Copper or titanium container large enough to contain the light bulb
  and fixture (opaque UV-resistant plastic is also fine)
01x Light bulb fixture with plug or battery powering

BUBBLER
01x 4" OD, 3.75" ID pipe size UV-resistant clear acrylic PVC
02x 4" PVC caps
01x 80mm/100mm stainless steel air diffusing stone (≤100microns)
12x Stainless steel scouring pads or finer steel wool
01x 1.5 - 2L of purified water
01x 0.495" ID, 0.625" OD, 0.065" Valve

VAJRA
04x 4" OD, 0.125" thick 304 alloy steel hemispheres
04x 3" OD, 0.125" thick 304 alloy steel hemispheres
04x 2" OD, 0.125" thick 304 alloy steel hemispheres
03x 3", 0.995" ID, 1.125" OD, 0.065" wall 304 alloy steel pipes
01x 4", 0.495" ID, 0.625" OD, 0.065" wall 304 alloy steel pipe
02x 2", 0.495" ID, 0.625" OD, 0.065" wall 304 alloy steel pipes
04x 1", 0.25" diameter 304 alloy steel dowel pins

ENGINE
01x Custom-milled aluminum extra-intake connector piece
  [see PHOTOS, page 38]
01x Honda EG 3600/4500/5500 CL motor (270cc, 337cc)
  *Other tested engines: 224cc (Phil Dubois), 308cc (Nykta Vovk),
  389cc and 439cc (NCSU London)

GAS ANALYZER PROBE

Although we're converting engine exhaust into breathable air, we recommend using a gas analyzer probe to test the exhaust's chemical content to make sure it is indeed breathable before you take a big whiff. Carbon monoxide (CO) can be lethal.

This is an emerging, rediscovered science so you, like others around the world who are building these machines as experiments, should collect your test results to share. Below you'll find a chart of real analyzer readings people have logged in their home tests [4]. (Also, see PHOTOS, page 39)

AVERAGE BASELINE EXHAUST READINGS	AVERAGE RETROFITTED EXHAUST READINGS
USING A SMALL HONDA MOTOR WITHOUT THE THUNDERSTORM GENERATOR	USING A SMALL HONDA MOTOR WITH THE THUNDERSTORM GENERATOR
CARBON DIOXIDE (CO2) = 9.3%	CARBON DIOXIDE (CO2) = 1.1%
CARBON MONOXIDE (CO) = 5.48%	CARBON MONOXIDE (CO) = 0%
OXYGEN (O) = 4.66%	OXYGEN (O) = 19.31%
HYDROCARBON (HC) = 119ppm	HYDROCARBON (HC) = 37ppm
NITROUS OXIDE (NOX) = 27ppm	NITROUS OXIDE (NOX) = 13ppm

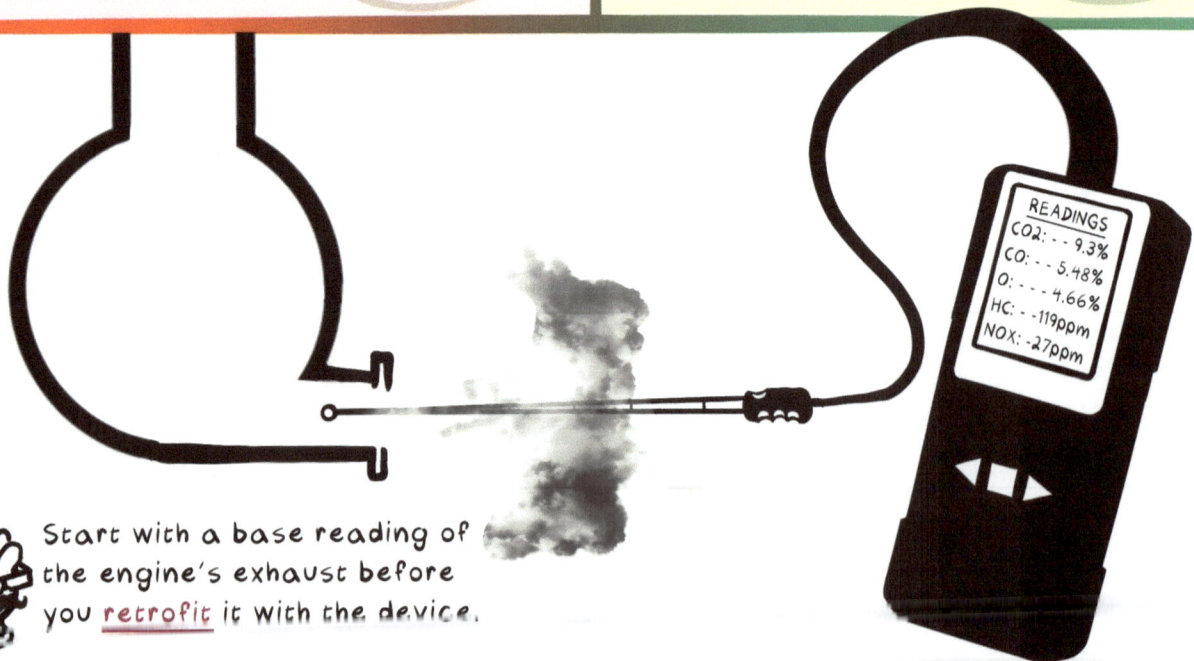

READINGS
CO2: - - 9.3%
CO: - - 5.48%
O: - - - 4.66%
HC: - - 119ppm
NOX: -27ppm

Start with a base reading of the engine's exhaust before you retrofit it with the device.

SECTION KEY

The following guide showcases how information will be displayed for each stage of the system's operation: Ionization Chamber, Bubbler, Vajra: Inner Chamber, Engine, then back through the Vajra: Outer Chamber.

CURRENT STAGE

☝ Quick Notes

SCIENCE!!!

WHAT
CONTINUES TO
NEXT STAGE

? **WHAT'S GOING ON HERE**
This section explains what's pictured.

💡 **FURTHER INFORMATION**
This section expands upon elements of the process.

🛍 **SUPPLIES**
This section lists materials and their specifics.

📐 **BUILD INSTRUCTIONS**
This section instructs how to assemble the SUPPLIES.

📄 **NOTES**
This section lists additional thoughts.

02x (FITTING) 0.5" Right-angle
BUILD INSTRI
01) Drill a 0.5" hole at the b
the PIPES and FITTINGS

SUPPLIES features truncated descriptions formatted "(ITEM)" that serve as keys for the BUILD INSTRUCTIONS.

Words highlighted like this can be further explained in the GLOSSARY on page 50.

"[#]" refers to the citation number for the BIBLIOGRAPHY on page 56, which contains links to source videos.

IONIZATION CHAMBER

The UV light bulb required can generally be found in the reptile section of most pet stores.

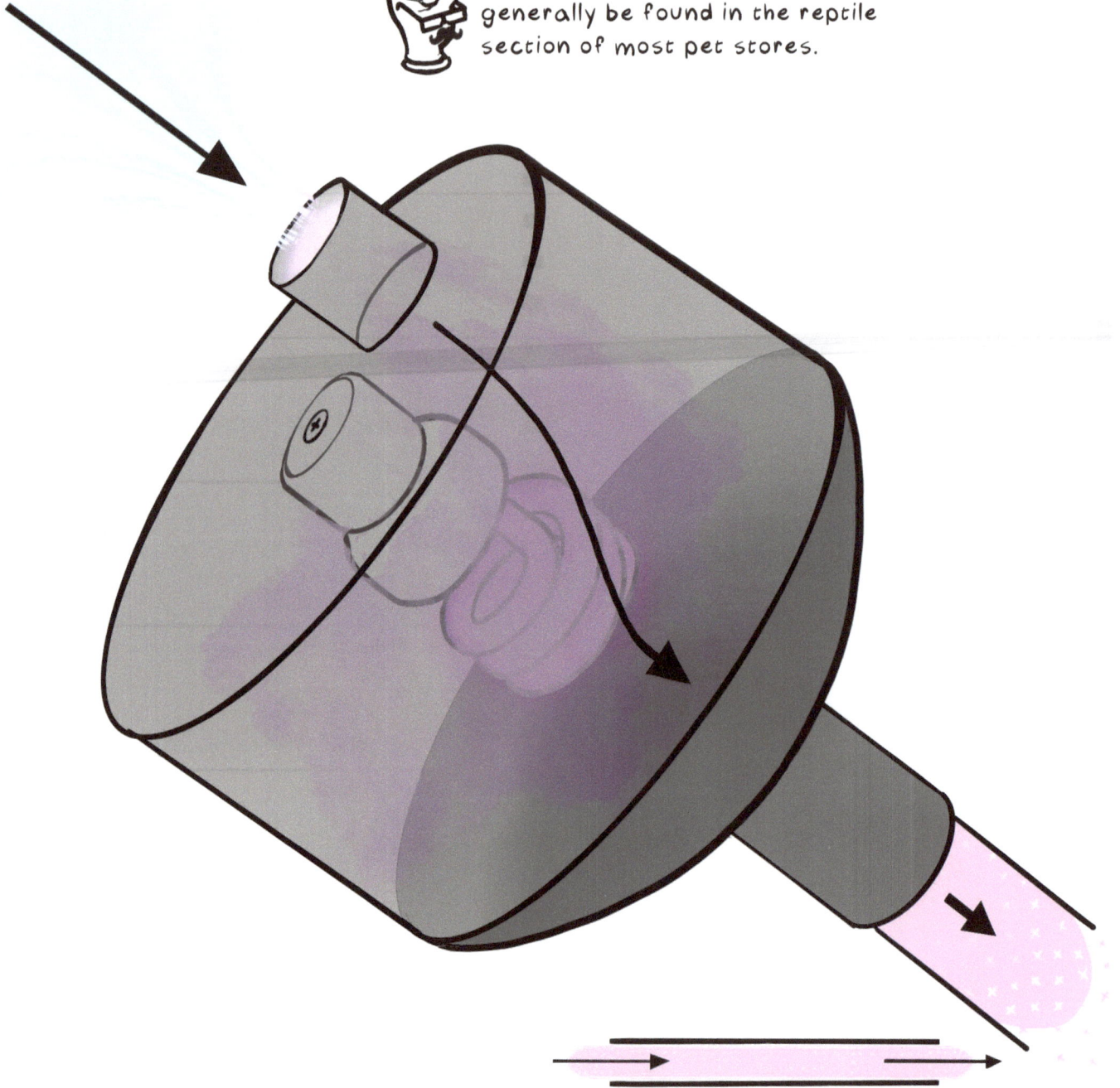

IONIZED AIR
CONTINUES TO
BUBBLER

? WHAT'S GOING ON HERE

Air gets vacuumed by the motion of the pistons in the motor into this container that houses an ultraviolet light bulb that ionizes and positively charges the atoms.

FURTHER INFORMATION

IONIZATION occurs when ultraviolet (UV) light gives atoms a powerful energy boost, exciting electrons to break free from their orbits. The process of atoms losing their electrons and gaining a positive charge ("exciting" the atom) transforms them into ions. Our sun is doing this naturally all around us, but we're talking about a concentrated bulb. Make sure the UV light bulb is kept in UV-shielding material. Don't look straight at the bulb (just like you wouldn't stare at the sun).

SUPPLIES

01x (BULB) E27 UVC sanitizer ozone-free, 25W 254nm wavelength light bulb
01x (FIXTURE) Light bulb fixture with plug or battery powering
01x (CONTAINER) Copper or titanium container large enough to house the BULB and FIXTURE (opaque UV-resistant plastic is also fine)
03x (PIPE) 0.495" ID, 0.625" OD, 0.065" Tungsten pipes (can also use copper or titanium)
02x (FITTING) 0.5" Right-angle air conditioner fitting

BUILD INSTRUCTIONS

01) Drill a 0.5" hole at the bottom of the CONTAINER, then connect the PIPES and FITTINGS to create a pathway leading to the Bubbler.
02) On the opposite side of the hole you just created, fasten the light bulb FIXTURE to the inside of the CONTAINER.
03) On the same side as the FIXTURE, create another hole to act as an inlet for outside air. You can feed the wiring from the FIXTURE through here, if you'd like. ☑ IONIZATION CHAMBER!

NOTES

This device has shown to be more efficient in humid climates because Malcolm originally built and tested his machine in the Maldives.

BUBBLER

The porous stone required can generally be found in the fish section of most pet stores.

100 micron-sized ionized air bubbles expand

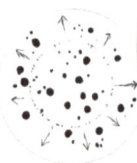

and collapse from the quick repetition of suction from the pistons of the engine. The atoms are shifted around, which creates an electric spark inside the bubbles [see ATOMIC BREAKDOWN, page 40]. This atomic but high heat event facilitates the transition from gas to plasma, the next state of matter (just like how adding different levels of heat to ice transitions its state to liquid, then gas).

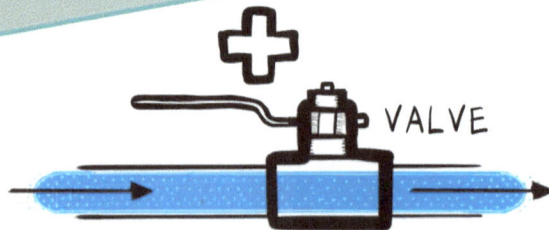

VALVE

IONIZED PLASMA CONTINUES TO VAJRA

? WHAT'S GOING ON HERE

The ionized air gets suctioned into this chamber, which houses a porous stone, water, and steel wool. The air is pulled through the porous stone, into the water, and through the steel wool in order to create billions of 100 micron-sized bubbles. The suction's impact on these specifically sized bubbles converts the gas to cold, wet plasmoids.

FURTHER INFORMATION

PLASMOIDS are consistent, blob-like structures of ionized ("excited") vapor with magnetic fields that can store high amounts of energy. If you've ever felt a static shock, you've encountered a plasmoid.

SUPPLIES

01x (PVC) 4" OD, 3.75" ID nominal pipe length UV-resistant clear acrylic PVC
02x (CAPS) 4" PVC caps
01x (STONE) 80mm/100mm stainless steel air diffusing stone (≤100microns)
12x (WOOL) Stainless steel scouring pads or finer steel wool
01x (WATER) 1.5 - 2L of pure water
03x (PIPE) 0.495" ID, 0.625" OD, 0.065" Tungsten pipes (Can also use copper or titanium)
02x (FITTING) 0.5" Right-angle air conditioner fittings
01x (VALVE) 0.495" ID, 0.625" OD, 0.065" Valve

BUILD INSTRUCTIONS

01) Fasten a single PIPE to one CAP.
02) Connect the IONIZATION CHAMBER PIPE to the other CAP. Fasten with sealant on both the outside and inside. Connect the PVC and seal.
03) Place the STONE 1" inside the PVC and fasten with sealant. Dry.
04) Insert the WOOL inside the longer end of the PVC next to the STONE.
05) Pour the WATER inside the WOOL end of the PVC.
06) Connect the remaining PIPED CAP to the end of the PVC with sealant. Connect the VALVE to this PIPE, then connect the final PIPE to the VALVE. ☑ **BUBBLER!**

NOTES

Water may need to be topped off at times due to plasmoid creation.

VAJRA: INNER CHAMBER

The resonance from the instrument affects the resonance of the molecules within it.

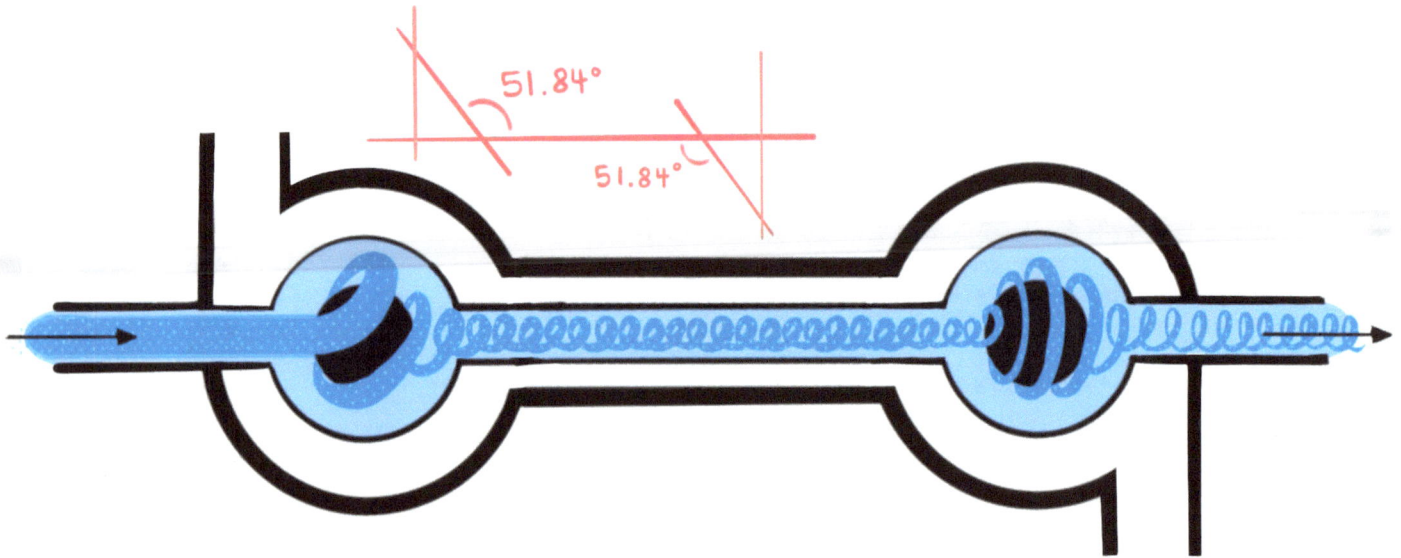

51.84°

51.84°

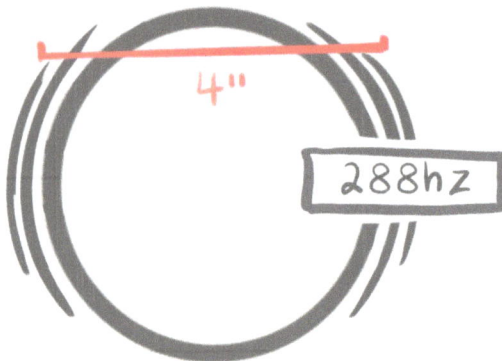

4"

288hz

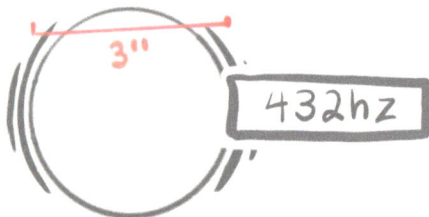

3"

432hz

2"

648hz

HARMONIC FREQUENCY CHART					
1 →x3	3	9	27	81	24
x2↓ x1.5					
2	6	18	54	162	48
4	12	36	108	324	97
8	24	72	216	648	194
16	48	144	432	1296	38
32	96	288	864	2592	77
64	192	576	1728	5184	155

IONIZED PLASMA CONTINUES TO ENGINE INTAKE

? WHAT'S GOING ON HERE

The cold, ionized plasma forms a vortex spiral with the help of precisely measured spheres to the ratio of 4:3:2. It continues down the Inner Chamber of the Vajra on its way to the Engine's Intake, just after the carburetor.

FURTHER INFORMATION

A vortex naturally forms because of the careful measurements of the spheres, located inside one another, with diameters of 4", 3", and 2". Malcolm Bendall found these measurements through his research into sacred geometry that correlates harmonic frequencies and atomic behavior with celestial measurements (like the sun's radius of 432,000 miles). [7]

SUPPLIES

04x (4" HALF) 4" OD, 0.125" thick 304 alloy steel hemispheres
04x (3" HALF) 3" OD, 0.125" thick 304 alloy steel hemispheres
04x (2" HALF) 2" OD, 0.125" thick 304 alloy steel hemispheres
03x (3" PIPE) 3", 0.995" ID, 1.125" OD, 0.065" wall 304 alloy steel pipes
01x (4" PIPE) 4", 0.495" ID, 0.625" OD, 0.065" wall 304 alloy steel pipe
02x (2" PIPE) 2", 0.495" ID, 0.625" OD, 0.065" wall 304 alloy steel pipes
04x (PIN) 1", 0.25" diameter 304 alloy steel dowel pins
SUPPLIES AFTER BUILDING THE VAJRA
01x (PIPE) 0.495" ID, 0.625" OD, 0.065" Tungsten pipe (can also use copper or titanium)
01x (FITTING) 0.5" Right-angle air conditioner fitting

BUILD INSTRUCTIONS

See VAJRA BUILD INSTRUCTIONS, page 28.
INSTRUCTIONS AFTER BUILDING THE VAJRA
01) Connect the PIPE coming in from the VALVE to either side of the 2" PIPE of the Inner Chamber. Use a FITTING and PIPE if needed.

NOTES

Bob Greenyer's build notes are available online. [8]

VAJRA

BUILD INSTRUCTIONS

(See previous section for SUPPLIES key. PRO TIP: Layout and label all supplies beforehand.)

01) Pair up the 4" HALVES so they form spheres and temporarily secure them with strong tape. For both, centered where the two halves meet, drill a 0.5" hole. Drill a 1" hole on the opposite ends.

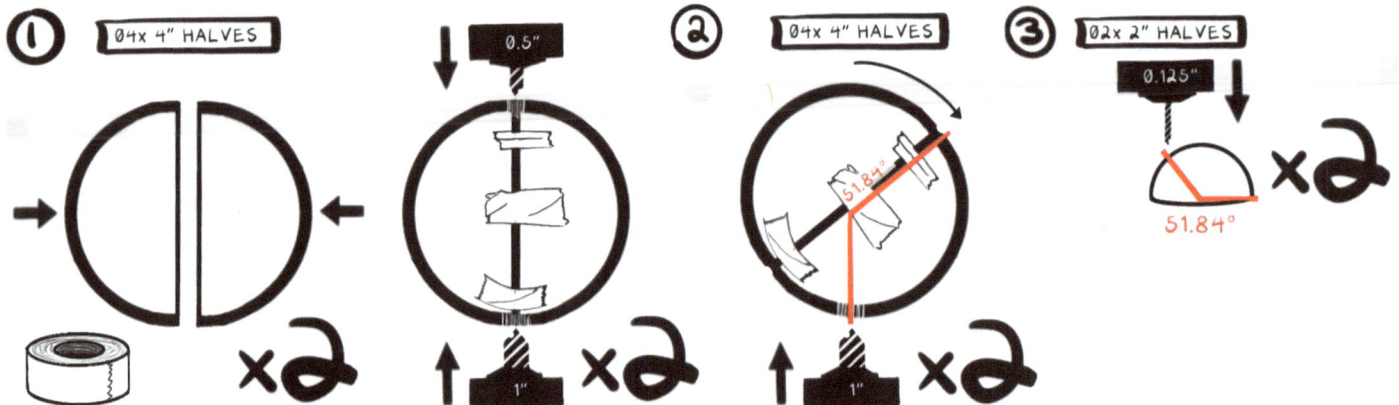

02) Tilt the 4" SPHERES exactly ∠51.84° and drill another 1" hole. Now remove the tape to transform them back into 4" HALVES.

03) Drill a 0.125" pressure release hole into two of the 2" HALVES at ∠51.84°. Pair these with the others.

04) Seperately weld a 3" PIPE to a 51.84° hole on the outside of a 4" HALF.

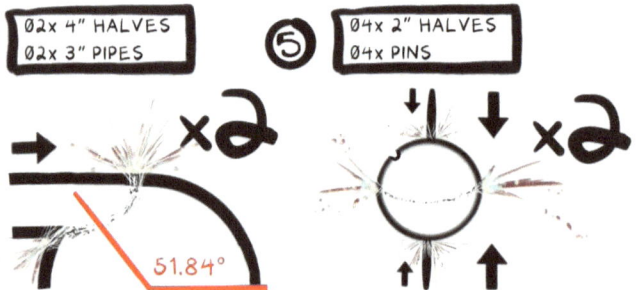

05) Next, weld the paired 2" HALVES together to form two 2" SPHERES. Weld the tip of a PIN on polar opposite sides of the 2" SPHERES.

06) We're now going to weld the PINS to the inside edges of two 3" HALVES. Place the 2" SPHERES inside the 3" HALVES and weld them so that only the PINS are touching their respective HALF. Weld the other 3" HALVES over their respective tops, making a pair of 3" SPHERES.

28

6 02x 2" SPHERES
04x 3" HALVES

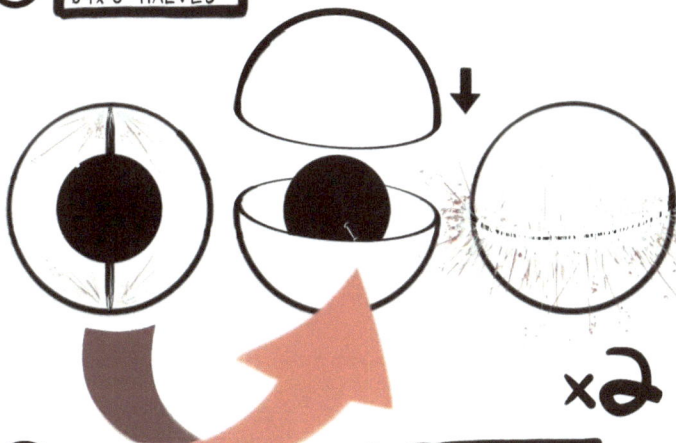

x2

7 02x 3" SPHERES

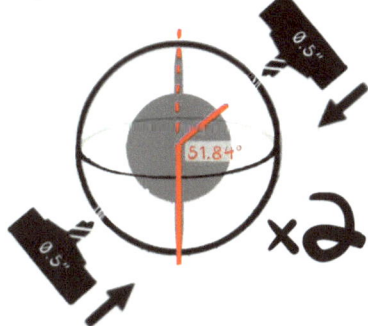

51.84°

0.5"

0.5"

x2

8 02x 3" SPHERES
02x 2" PIPES

x2

07) Picture for a moment that an imaginary line is connecting the two PINS. One of the PINS is 90° and the opposite PIN is 0°. Drill a 0.5" hole at ∠51.84°, and a second hole on the polar opposite end.

08) Weld a 2" PIPE to one hole per sphere.

09) Now is a good time to pressure test (page 36) your work to make sure there are no air leaks in your welding. There's no going back after we weld these 3" SPHERES inside the 4" HALF housing.

10) Weld one end of the 4" PIPE to the 0.5" hole on one of the 3" SPHERES. Insert the 4" PIPE through the remaining 3" PIPE, then weld that end of the 4" PIPE to the other 3" SPHERE's 0.5" hole.

10 01x 4" PIPE
01x 3" PIPE
02x 3" SPHERES

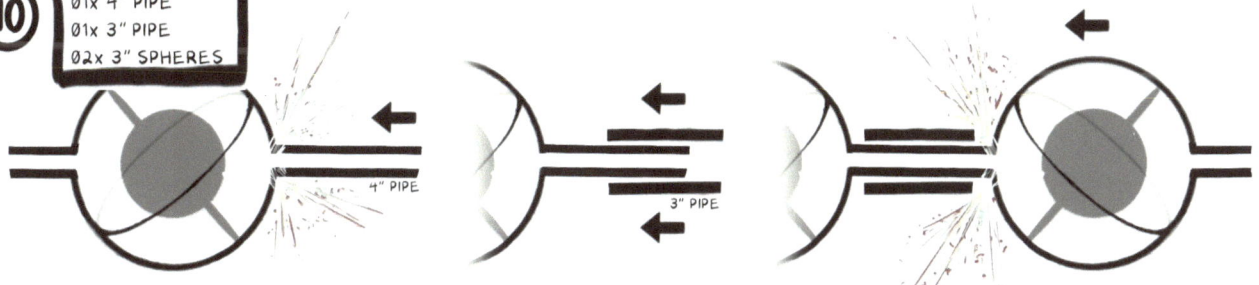

4" PIPE

3" PIPE

11) Finally, we're going to seal the 3" SPHERES inside of the 4" HALF housing. Assemble the 4" HALVES to seamlessly and securely click onto the INNER CHAMBER. Make sure both ends have one 3" PIPE facing the opposite direction. Weld the 4" HALVES to the 2" and 3" PIPES of the INNER CHAMBER, then weld the 4" HALVES together. ☑ **VAJRA**

12) Do a final pressure test on your welding to make sure you're locked up air tight all around.

SOAPY WATER TEST on page 36.
Need help? Sourced videos linked in BIBLIOGRAPHY, page 56.

11 04x 4" HALVES
01x INNER CHAMBER

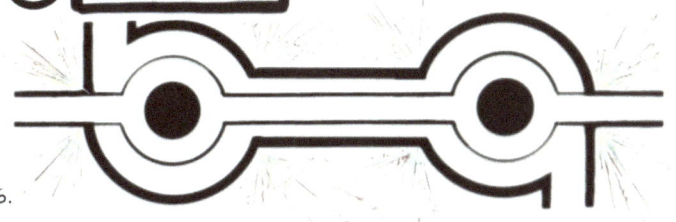

ENGINE* INTAKE/EXHAUST

*Engines may vary.

EXHAUST

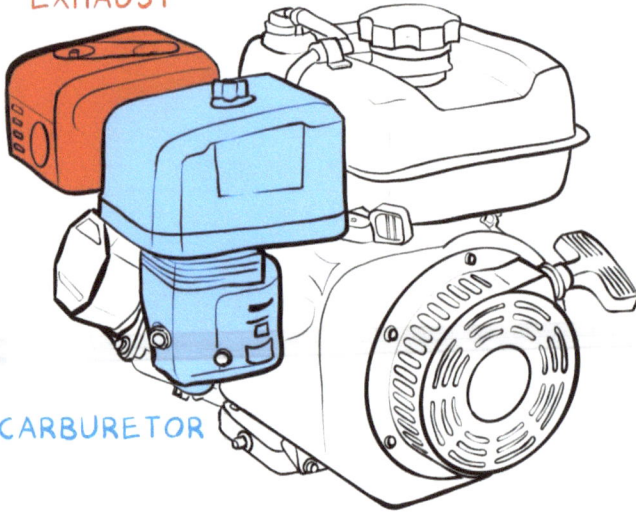

CARBURETOR

Methane gas is the cleanest gas to use.

Cleans excess carbon out of the entire system. Say good-bye to oil changes!

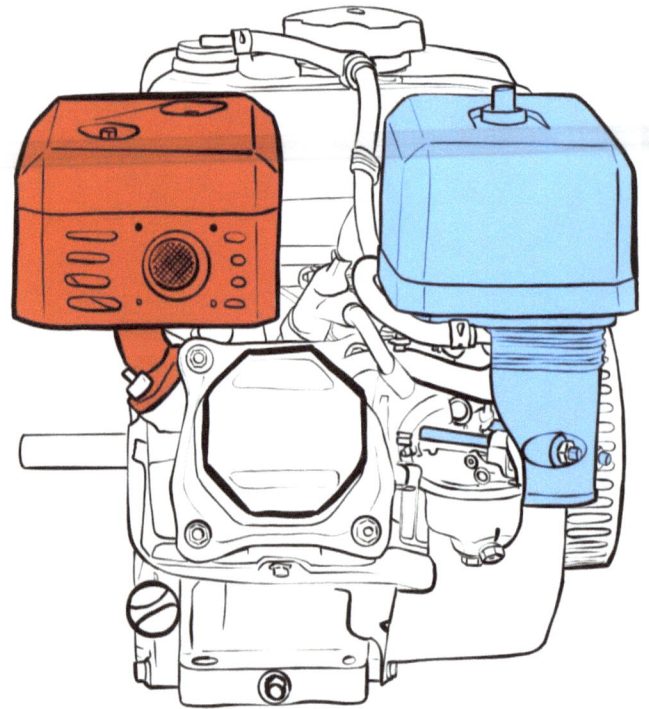

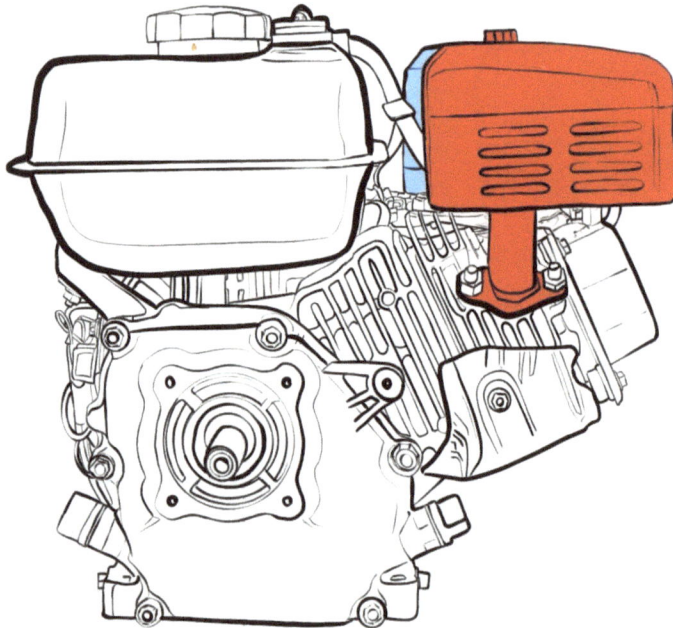

Honda Predator 6.6 HP (224cc)

TYPICAL ENGINE OPERATION EXPELS EXHAUST TO VAJRA

? WHAT'S GOING ON HERE

The cold plasmoids are injected straight into the engine's intake. You'll hear an audible change in the engine as its efficiency increases. The usual carbon-rich exhaust will exit the engine's tailpipe and be channeled into the Vajra's Outer Chamber for processing.

FURTHER INFORMATION

Carbon build-up and oil throughout the entire engine is cleaned from its interaction with plasmoids. The engine was observed to continue to operate at double efficiency for 8 years *after* the retrofit's removal [8]. The only explanation so far is Malcolm's claim, "The engine becomes drenched in plasmoids." [1]

SUPPLIES

01x (PIECE) Custom-milled aluminum connector piece (see PHOTOS, page 38)

01x (MOTOR) Honda EG 3600/4500/5500 CL motor (270cc, 337cc)

*Other tested engines: 224cc (Phil Dubois), 308cc (Nykta Vovk), 389cc and 439cc (NCSU London)

BUILD INSTRUCTIONS

01) The output of the Inner Chamber needs to be plumbed between the engine and the carburetor. This can be achieved by precision milling an aluminum block for this purpose. Recheck vacuum integrity after install.

02) The fuel tank and injector are moved to make room for the custom PIECE.

NOTES

ENGINE OPERATION (AFTER DEVICE IS INSTALLED):

01) Close the VALVE. [see Bubbler, page 24]

02) Turn the choke of the carburetor to ON/START.

03) Turn on the MOTOR per usual operation. Let it run for 30 seconds.

04) Open the VALVE to partially restrict the air flow in while calibrating the engine to run optimally with its new retrofit.

VAJRA: OUTER CHAMBER

Exhaust Temperature:
392°F (200°C)
Traveling Clockwise

Plasmoids Vapor Temperature:
-58°F (-50°C)
Traveling Counter-Clockwise

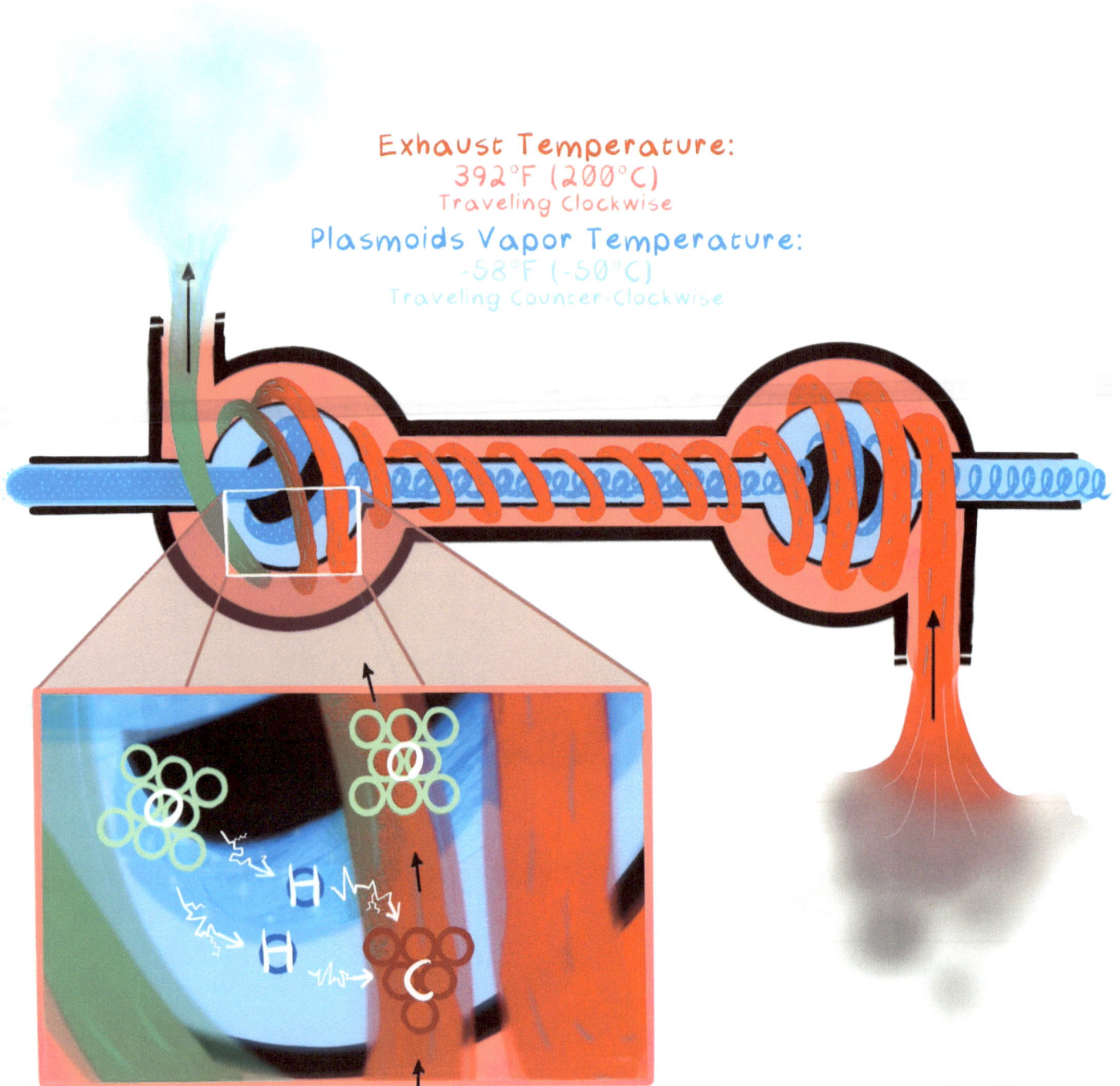

Protons are reconfigured from **carbon** and H_2O to create two **oxygen** atoms.

There is no radiation emitted by this system.

EXHAUST
CONVERTS TO
OXYGEN

? WHAT'S GOING ON HERE

The hot exhaust shoots counter-clockwise down the Outer Chamber of the Vajra and interacts atomically with the cold, clockwise-spinning ionized plasma through the Inner Chamber. It adds protons and electrons to the carbon molecule and converts it into clean, breathable oxygen. (see ATOMIC BREAKDOWN, page 40)

FURTHER INFORMATION

The loose but controlled state of cold ionized protons fusing atomically with hot carbon to convert the element into oxygen by bumping its atomic number from 6 to 8 illustrates true alchemy. The potential is staggering. The number of cars in the world is estimated to be around 1.48 billion. Imagine if we retrofitted every single one, dramatically reducing the amount of carbon gases in the atmosphere. Engines double in efficiency, no longer pollute the air we breathe, and don't require oil changes. The Thunderstorm Generator opens up a world of possibilities. Oh, and Nikola Tesla's dream of free energy for all? That's Malcolm's goal, too. And now we have the technology to make it a reality. (see SUPPORT, page 08)

SUPPLIES

This section mirrors the VAJRA: INNER CHAMBER: SUPPLIES on page 27.

BUILD INSTRUCTIONS

See VAJRA BUILD INSTRUCTIONS, page 28.

NOTES

After the engine warms up and the exhaust reaches its highest temperature, the reduction in carbon and production of oxygen are at peak performance. [8]

TROUBLESHOOTING

THINGS TO LOOK OUT FOR

—STATIC FLASHES OUTSIDE THE VAJRA:

If you see static sparks (plasmoid flashes) outside the Vajra, the sphere sizes might be incorrect (remember, we need a precise 4:3:2 ratio). Make sure your ratio sizes are exact because you want to contain these flashes.

—NOT WORKING AND THE VALVE IS ALL THE WAY OPEN:

The air flow needs to be dialed in so that the suction occurs just right. Remember to start with the valve closed. Adjust the air intake through the valve and carburetor. Check for leaks or plug-ups. You'll hear the engine improve when it's at optimal settings.

PERPLEXING RESULTS

—TEMPERATURE DIFFERENCE IN OUTER CHAMBER:

Tests conducted in London by George Lush, an aerospace annealing expert, measured the temperature of the engine's exhaust pipe leading into the outer sphere of the Vajra to be around 195°C (383°F). Shockingly, the temperature dropped by 100°C on the opposite side of the sphere, dramatically alternating between temperatures as it swirled around the chamber. Thermal imaging shows a solid line between the two temperatures of the spinning (1000 rpm) exhaust. According to the laws of thermodynamics, this phenomenon should be impossible, but it was observed again by many enthusiasts at the Cosmic Summit 2024.

—IONIZATION CHAMBER AND BUBBLER ARE A PRIMER:

During those same demonstrations, the Ionization Chamber and Bubbler were discovered to be damaged, yet even while operating without these steps the Thunderstorm Generator achieved similar results:

- CO_2 reduced from 6% to 0%
- CO reduced from 7.57% to 0.02%
- HC reduced from 1,138ppm to 162ppm
- NO_x reduced from 9.5ppm to 2ppm
- O_2 output increased from 4.55% to 20.86%

Malcolm stated that the residual plasmoids in the Vajra maintained their charge from the London tests. [8] Wild!

SOAPY WATER TEST

How to test that your welding is airtight using soap **and** water!

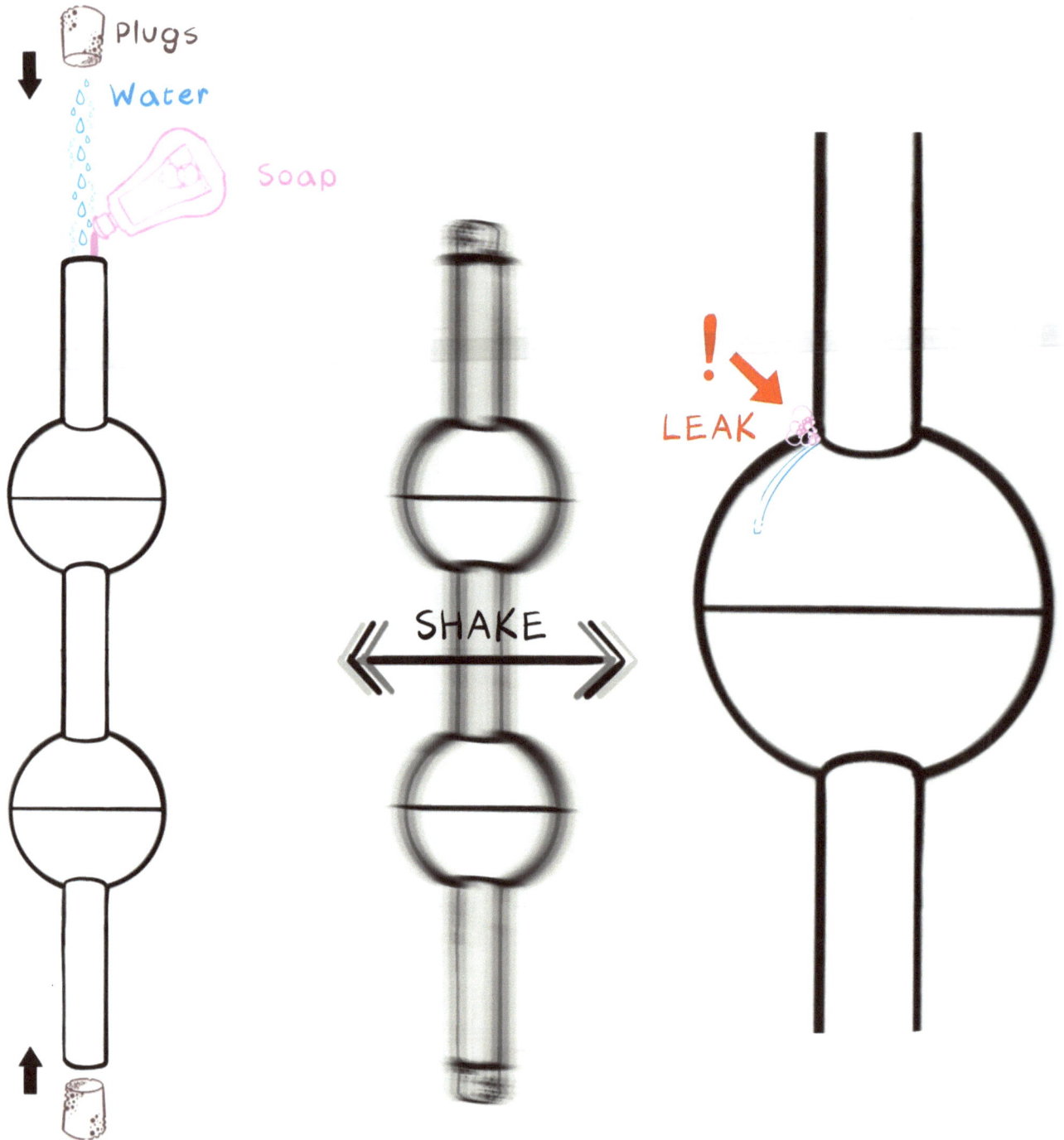

Plugs

Water

Soap

SHAKE

! LEAK

TO FIX: Re-weld or use a sealant. Maintaining an airtight chamber and geometric integrity is key.

TEST RESULTS / NOTES:

Figure XX. Aluminum intake air inlet to be placed between engine and carburetor. The bore is a 1" diameter. The connection is threaded ½" NPT. Fabricated by Robert Hutchings.

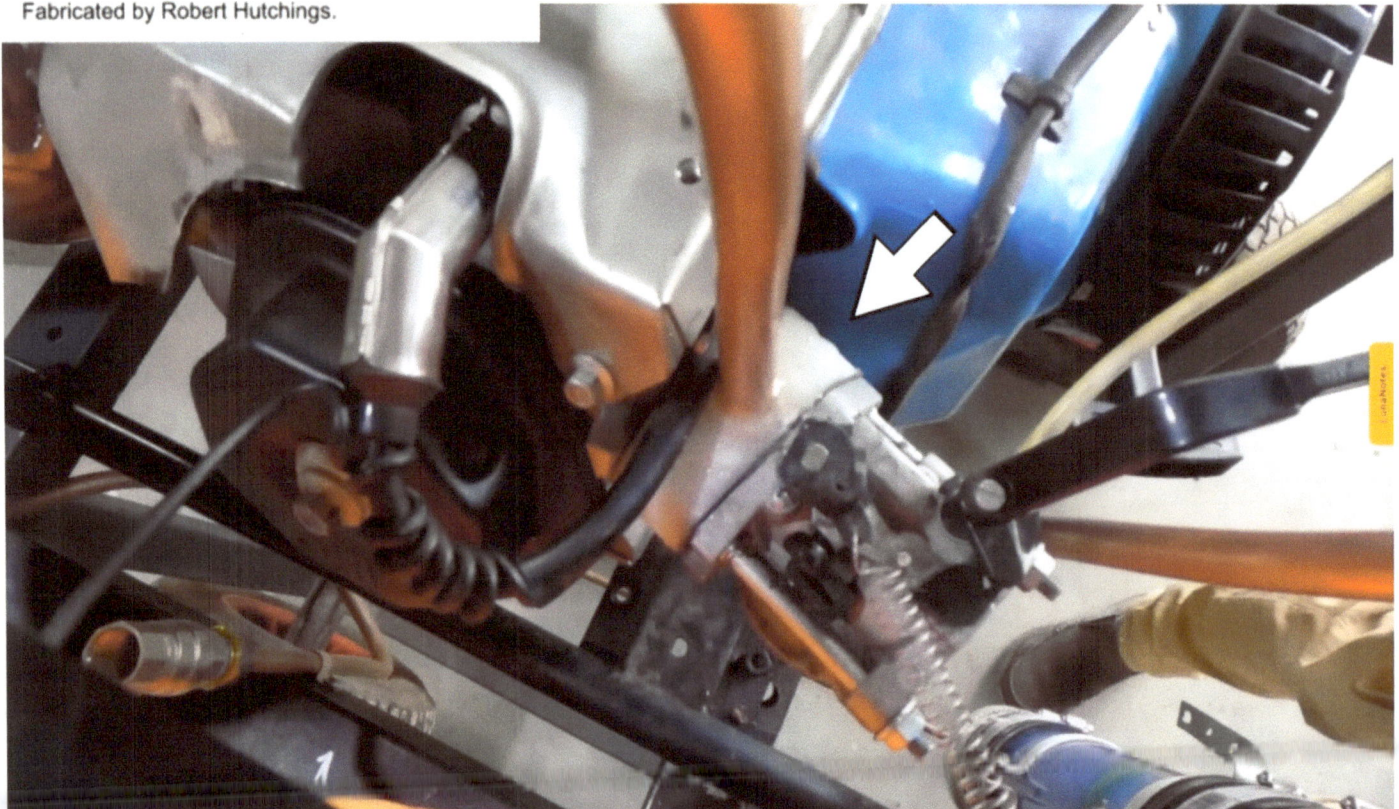

On this page we'll look at the custom-milled aluminum connector piece mentioned in ENGINE INTAKE/EXHAUST: SUPPLIES on page 31. On the following page are screenshots of different gas analyzer probe readings of the exhaust from engines with different Thunderstorm Generators retrofitted on them by various testers. As you can see, results vary only slightly with different build types. This means the fundamental reaction is taking place in all design tweaks.

LEFT: This is the custom-milled aluminum connector piece displayed in Bob Greenyer's "THOR" paper. [08]

BELOW: This picture shows the custom-milled aluminum connector piece in place between the engine's intake and carburetor. [01]

The original design
of the Vajra. [Ø7]

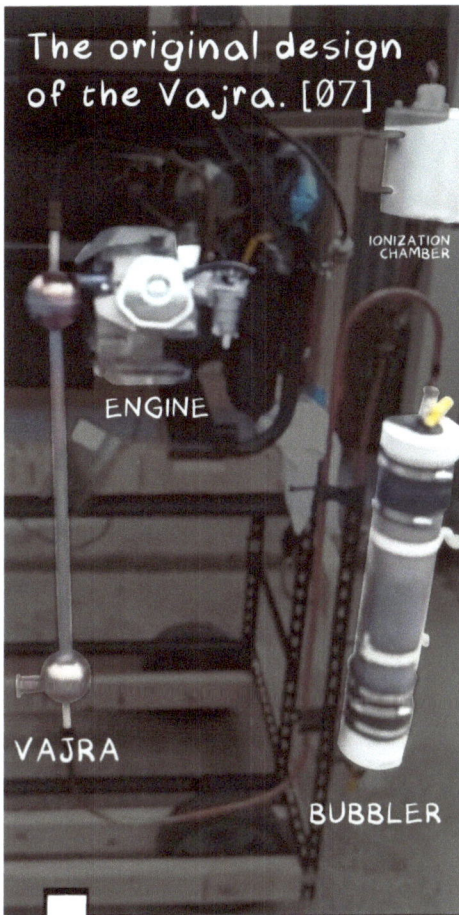

IONIZATION
CHAMBER

ENGINE

VAJRA

BUBBLER

RESULTS
O2: 20.50%
CO2: 0.30%

A design where the
inner chamber
exits the side of
the sphere. [Ø4]

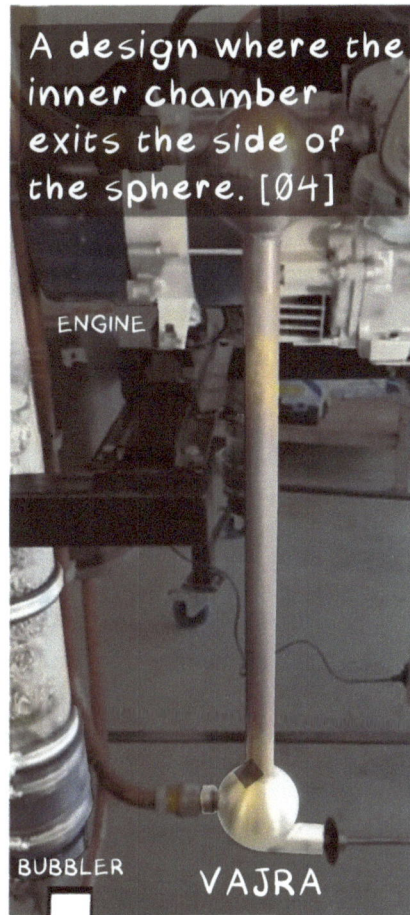

ENGINE

BUBBLER   VAJRA

RESULTS
O2: 17.20%
CO2: 0.00%

A shortened design.
[Ø7]

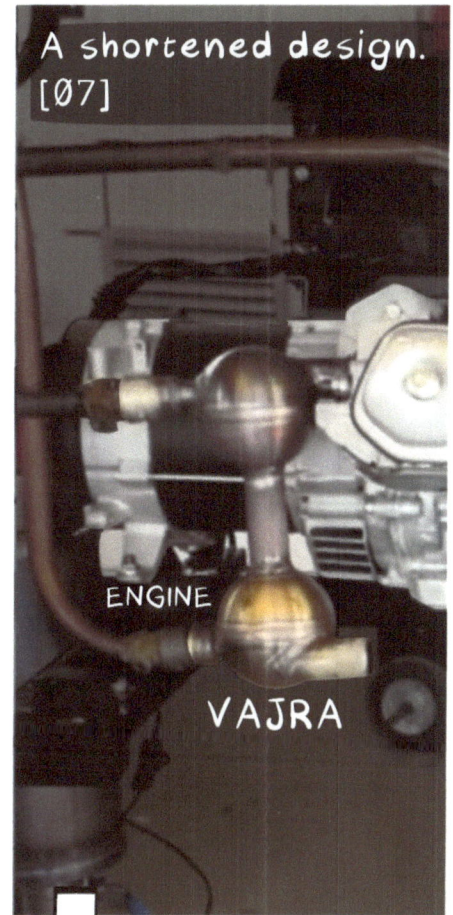

ENGINE

VAJRA

RESULTS
O2: 18.79%
CO2: 1.30%

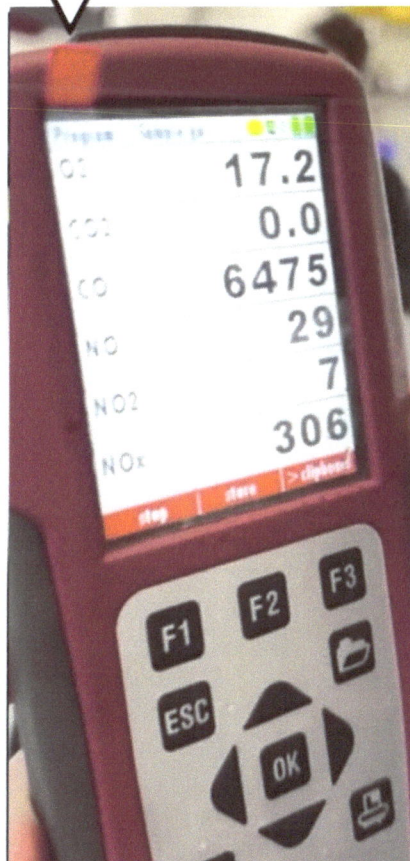

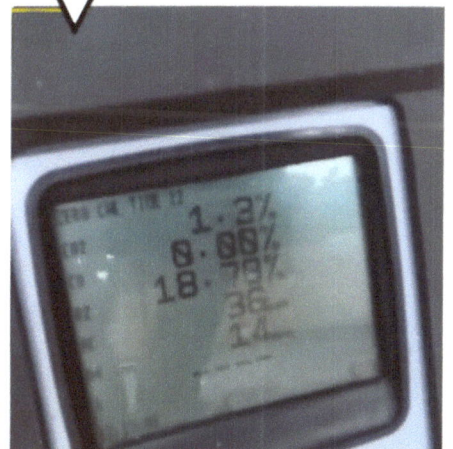

ABOVE: Here are three
different results showing
the increase in oxygen
and decrease in carbon
dioxide based on varied
early designs. Keep in
mind that results
fluctuate by a few points
while using a gas
analyzer probe.

ATOMIC BREAKDOWN

In this section we will describe what is taking place inside the Bubbler (page 24) and the Vajra (page 32).

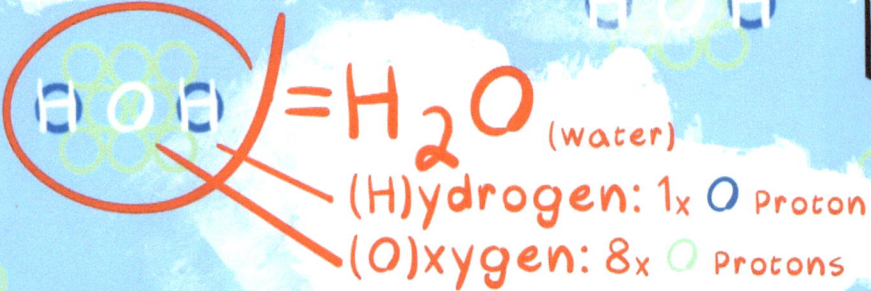

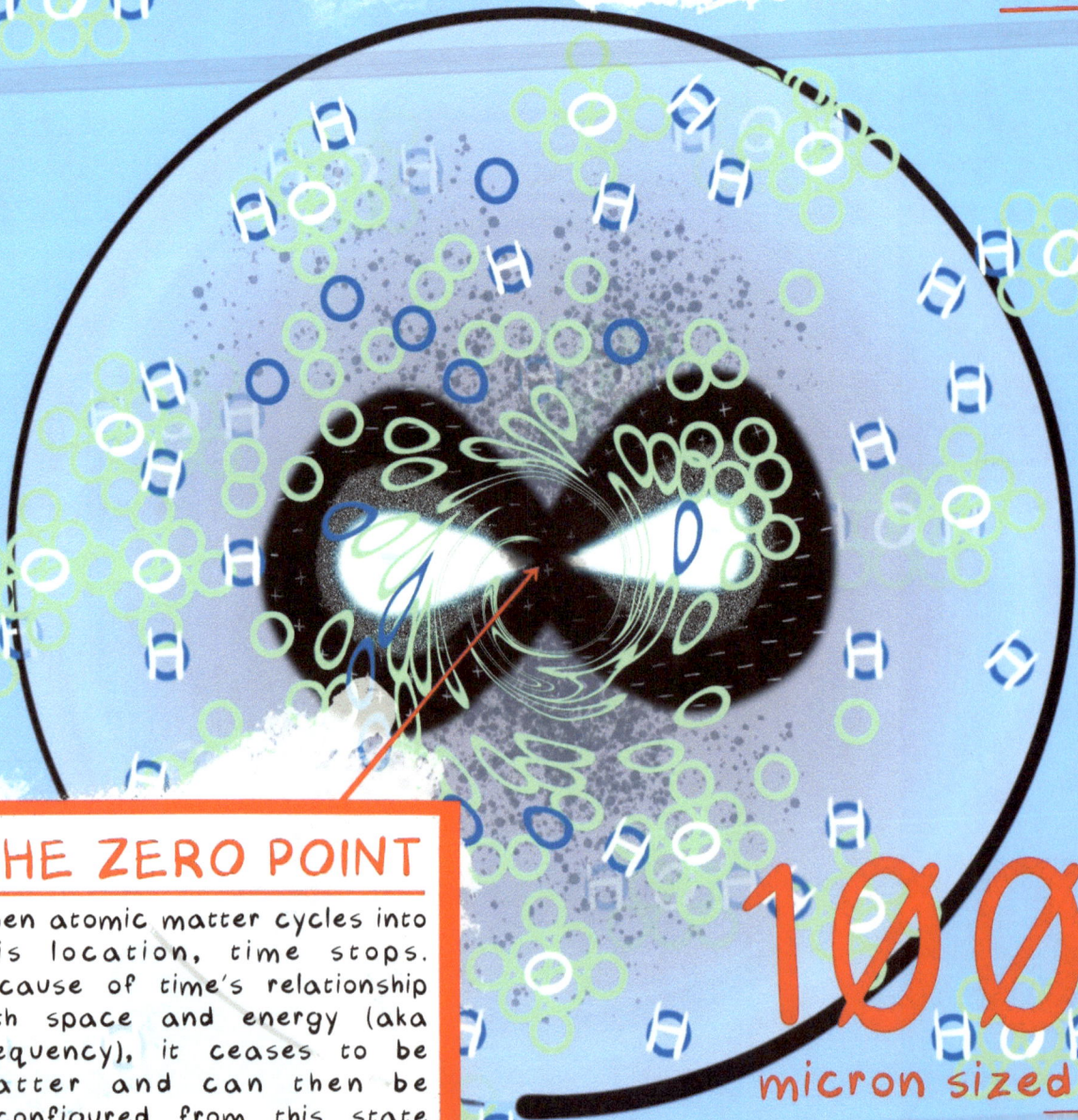

BUBBLER

$= H_2O$ (water)

(H)ydrogen: $1 \times$ ○ Proton
(O)xygen: $8 \times$ ○ Protons

THE ZERO POINT

When atomic matter cycles into this location, time stops. Because of time's relationship with space and energy (aka frequency), it ceases to be matter and can then be reconfigured from this state into other forms (see page 42)

100 micron sized

When The Elements Inside The Tiny Bubbles collapse and cause a static electric shock (100,000 watt electron volts), the heat transitions the surrounding molecules into plasma. The water (H_2O) surrounding the bubbles gets stripped down to its elements: two (H)ydrogens and one (O)xygen. As you get closer to the center of the static shock, the surrounding elements get further reduced to their elemental building blocks: protons and electrons. These are stored within the magnetic field of energy in the plasma, only to be redistributed later using the mechanics of thunderstorms.

Thunderstorms Occur when a warm body of air comes into contact with a cold body of air. Cold air is heavier and wants to sink while hotter air wants to rise. The giant, heavy, moisture-retaining storm clouds create a slight vacuum underneath them as the pressure per square inch of air below rises. Above, when these two bodies of differing air temperatures barrel into one another, the atomic friction causes static electric shocks (lightning), much like what happens in our bubbles, though on a much smaller scale.

"AS ABOVE, SO BELOW." - *Hermes Trismegistus 200-800 CE*

The Cold, Plasma Soup of hydrogen (1 proton) and oxygen (8 protons) can pass through the metal of the Inner Chamber to reach the carbon (6 protons) because, at the level of atomic interactions, there is sooo much space between atoms (WELL over the 100 microns allotted to them in their bubbles).

Inside The Vajra, when the cold, counter-clockwise spinning, hydrogen-rich, ionized plasma interacts with the hot, clockwise spinning, carbon exhaust they cause an interaction with one another. Water's two hydrogen protons latch onto the carbon protons and make oxygen. This newly formed oxygen, combined with the oxygen left over from the breakdown of H_2O and CO_2, creates an exhaust of ~20% oxygen, and other clean noble gases.

A QUICK ASIDE
—ON—
SPACETIME

In Our Universe, everything moves and pulses with energy, measured as frequency—how often something happens over a set period of time, in cycles per second (Hz). Imagine a diving board springing back and forth, each movement a measure of its distance over time. Since we perceive time as always moving forward and everything vibrates, these two measurements are interconnected in spacetime. This connection between distance and time is fundamental: time influences how we perceive distances and vice versa. You can't travel a distance without time. They are so inseparable, they become the same thing, but we perceive them as separate.

All Matter And Phenomena Vibrate with unique frequencies. From colors to earthquakes and tastes to elements, everything has its own vibrational signature. Scientists measure these frequencies to create models that predict behavior on different scales, much like musical notes arranged in octaves and fifths. Every sensor in our body—eyes, ears, tongue, etc.—detects specific frequencies like light, sound, or taste, and sends that information to our brain, forming our sensory experience, shaping our reality.

Consider A Flute producing sound when air passes through it. The air vibrates at a frequency our ears detect, allowing us to hear sound. The flute's shape represents space, while the passage of air through it represents time. The vibrations (frequency) in the air create sound, similar to how vibrations in spacetime manifest as matter. Sound can't exist without the relationship between the flute and air, just as matter and energy are intertwined with spacetime.

When Matter (A Frequency) encounters a zone where time stands still, like in a theoretical state of zero-point energy, it ceases to move or vibrate. When time resumes, it's like tuning a flute to a new key—you can create something entirely different.

PERIODIC TABLES

That's right—table_s! Throughout history, there have been multiple ways to organize the elements. The most commonly seen is Dmitri Mendeleev's "Periodic Table of Elements" from 1869, which organizes elements by their atomic weight, number of protons, and chemical properties. In 1926, Walter Russell released his "Periodic Chart of Elements" in his book The Universal One proposing that elements and their Isotopes vibrate in a musical pattern akin to octaves and perfect fifths, arranging elements so that

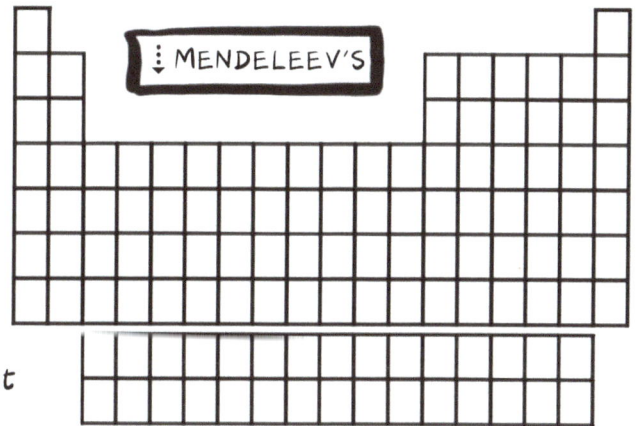

their atomic structures mirror these harmonic intervals. In 2022, Malcolm Bendall introduced his "Plasmoid Unification Model," which focuses on the role of protium (a hydrogen isotope) and its various energy states as they change with frequency, highlighting the resonant and interconnected nature of elements.

In the following pages we'll display these different tables with descriptions and keys on how to read them. Maybe it'll even inspire some creative insights of your own!

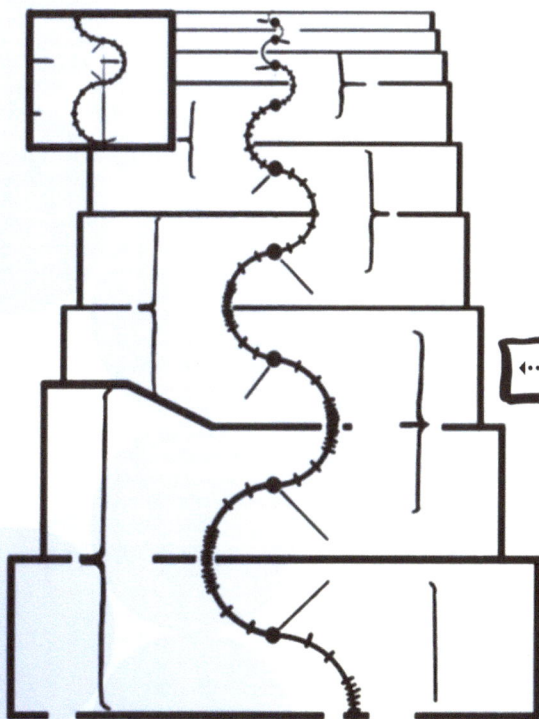

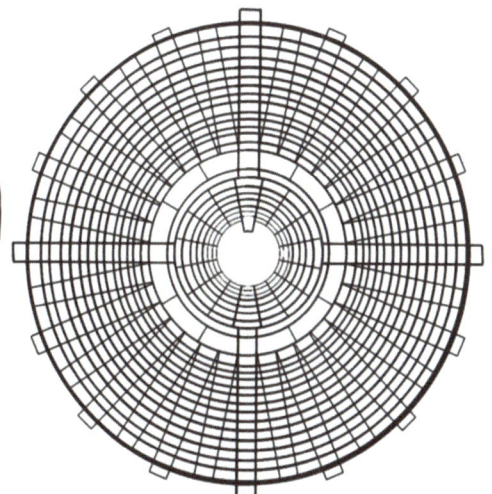

Periodic Table

1	2	3	4	5	6	7	8	9	10	11	12	13	14	15	16	17	18
1 **H** Hydrogen 1.008																	2 **He** Helium 4.003
3 **Li** Lithium 6.941	4 **Be** Beryllium 9.012											5 **B** Boron 10.811	6 **C** Carbon 12.011	7 **N** Nitrogen 14.007	8 **O** Oxygen 15.999	9 **F** Fluorine 18.998	10 **Ne** Neon 20.180
11 **Na** Sodium 22.990	12 **Mg** Magnesium 24.305											13 **Al** Aluminum 26.982	14 **Si** Silicon 28.086	15 **P** Phosphorus 30.974	16 **S** Sulfur 32.066	17 **Cl** Chlorine 35.453	18 **Ar** Argon 39.948
19 **K** Potassium 39.098	20 **Ca** Calcium 40.078	21 **Sc** Scandium 44.956	22 **Ti** Titanium 47.867	23 **V** Vanadium 50.942	24 **Cr** Chromium 51.996	25 **Mn** Manganese 54.938	26 **Fe** Iron 55.845	27 **Co** Cobalt 58.933	28 **Ni** Nickel 58.693	29 **Cu** Copper 63.546	30 **Zn** Zinc 65.38	31 **Ga** Gallium 69.723	32 **Ge** Germanium 72.631	33 **As** Arsenic 74.922	34 **Se** Selenium 78.971	35 **Br** Bromine 79.904	36 **Kr** Krypton 84.798
37 **Rb** Rubidium 85.468	38 **Sr** Strontium 87.62	39 **Y** Yttrium 88.906	40 **Zr** Zirconium 91.224	41 **Nb** Niobium 92.906	42 **Mo** Molybdenum 95.95	43 **Tc** Technecium 98.907	44 **Ru** Ruthenium 101.07	45 **Rh** Rhodium 102.906	46 **Pd** Palladium 106.42	47 **Ag** Silver 107.868	48 **Cd** Cadmium 112.414	49 **In** Indium 114.818	50 **Sn** Tin 118.711	51 **Sb** Antimony 121.760	52 **Te** Tellurium 127.6	53 **I** Iodine 126.904	54 **Xe** Xenon 131.294
55 **Cs** Cesium 132.905	56 **Ba** Barium 137.328	57-71 **Lu** Lutecium 174.967	72 **Hf** Hafnium 178.49	73 **Ta** Tantalum 180.948	74 **W** Tungsten 183.84	75 **Re** Rhenium 186.207	76 **Os** Osmium 190.23	77 **Ir** Iridium 192.217	78 **Pt** Platinum 195.085	79 **Au** Gold 196.967	80 **Hg** Mercury 200.592	81 **Tl** Thallium 204.383	82 **Pb** Lead 207.2	83 **Bi** Bismuth 208.980	84 **Po** Polonium 208.982	85 **At** Astatine 209.987	86 **Rn** Radon 222.018
87 **Fr** Francium 123.020	88 **Ra** Radium 226.025	89-103 **Lr** Lawrencium [262]	104 **Rf** Rutherfordium [262]	105 **Db** Dubnium [261]	106 **Sg** Seaborgium [266]	107 **Bh** Bohrium [264]	108 **Hs** Hassium [269]	109 **Mt** Meitnerium [278]	110 **Ds** Darmstadium [281]	111 **Rg** Roentgenium [280]	112 **Cn** Copernicium [285]	113 **Nh** Nihonium [286]	114 **Fl** Flerovium [289]	115 **Mc** Moscovium [289]	116 **Lv** Livermorium [293]	117 **Ts** Tennessine [294]	118 **Og** Oganesson [294]

Lanthanides

57 **La** Lanthanum 138.905	58 **Ce** Cerium 140.116	59 **Pr** Praseodymium 140.908	60 **Nd** Neodymium 144.243	61 **Pm** Promethium 144.913	62 **Sm** Samarium 150.36	63 **Eu** Europium 151.964	64 **Gd** Gadolinium 157.25	65 **Tb** Terbium 158.925	66 **Dy** Dysprosium 162.500	67 **Ho** Holmium 164.930	68 **Er** Erbium 167.259	69 **Tm** Thulium 168.934	70 **Yb** Ytterbium 173.055	71 **Lu** Lutecium 174.967

Actinides

89 **Ac** Actinium 227.028	90 **Th** Thorium 232.038	91 **Pa** Protactinium 231.036	92 **U** Uranium 238.029	93 **Np** Neptunium 237.048	94 **Pu** Plutonium 244.064	95 **Am** Americium 243.061	96 **Cm** Curium 247.070	97 **Bk** Berkelium 247.070	98 **Cf** Californium 251.080	99 **Es** Einsteinium [254]	100 **Fm** Fermium 257.095	101 **Md** Mendelevium 258.1	102 **No** Nobelium 259.101	103 **Lr** Lawrencium [262]

Legend

@ Room Temp.

- GAS
- LIQUID
- SOLID

- Alkali Metal
- Alkaline Earth
- Transition Metal
- Basic Metal
- Metalloid
- Nonmetal
- Halogen
- Noble Gas
- Lanthanide
- Actinide

Group/Family

Atomic Number — **Symbol** — Name — Atomic Weight

Period

DMITRI MENDELEEV'S
PERIODIC TABLE OF ELEMENTS

Reading the periodic table involves understanding its structure and the information presented for each element:

1 → GROUP/FAMILY

1 → ATOMIC NUMBER (PROTONS)

PERIOD → 1

H → SYMBOL

Hydrogen → NAME

1.008 → ATOMIC WEIGHT

PERIOD: The horizontal rows are called periods and are numbered from 1 to 7. Elements in the same period have the same number of atomic orbitals. Moving across a period, each element has one more proton and is less metallic than the element before it.

GROUPS/FAMILIES: The vertical columns are called groups or families, numbered from 1 to 18. Elements in the same group have similar properties and the same number of electrons in their outer orbital (valence electrons).

ATOMIC NUMBER: Located at the top of the box, it indicates the number of protons in the nucleus.

SYMBOL: One or two letters that are a shorthand for the element's name.

NAME: The full name of the element.

ATOMIC WEIGHT: The average mass of atoms of the element (in atomic mass units - amu).

- -

The table is color-coded to categorize the elements into different types, such as:

ALKALI METALS: Group 1, highly reactive metals.

ALKALINE EARTH METALS: Group 2, also highly reactive metals.

TRANSITION METALS: Groups 3-12, metals with typical metallic properties.

BASIC METAL: Group 13, located after the transition metals, these elements have properties more like the heavier alkaline earth metals.

METALLOID: Positioned along the zig-zag line dividing metals and nonmetals, metalloids have properties intermediate between metals and nonmetals.

NONMETALS: Located mostly on the right side of the table, these are diverse elements with varying properties, typically not shiny and poor conductors of heat and electricity.

HALOGEN: Group 17, very reactive nonmetals.

NOBLE GASES: Group 18, nonreactive gaseous elements. Quite common.

LANTHANIDES: These are the 14 elements from Cerium to Lutetium, often displayed at the bottom of the periodic table to save space, and are used in various high-tech applications.

ACTINIDES: These are the 14 elements from Thorium to Lawrencium, also displayed at the bottom. They include radioactive elements and are often involved in nuclear applications.

- -

Also marked is the physical state of the element at room temperature: Gas, Liquid, or Solid.

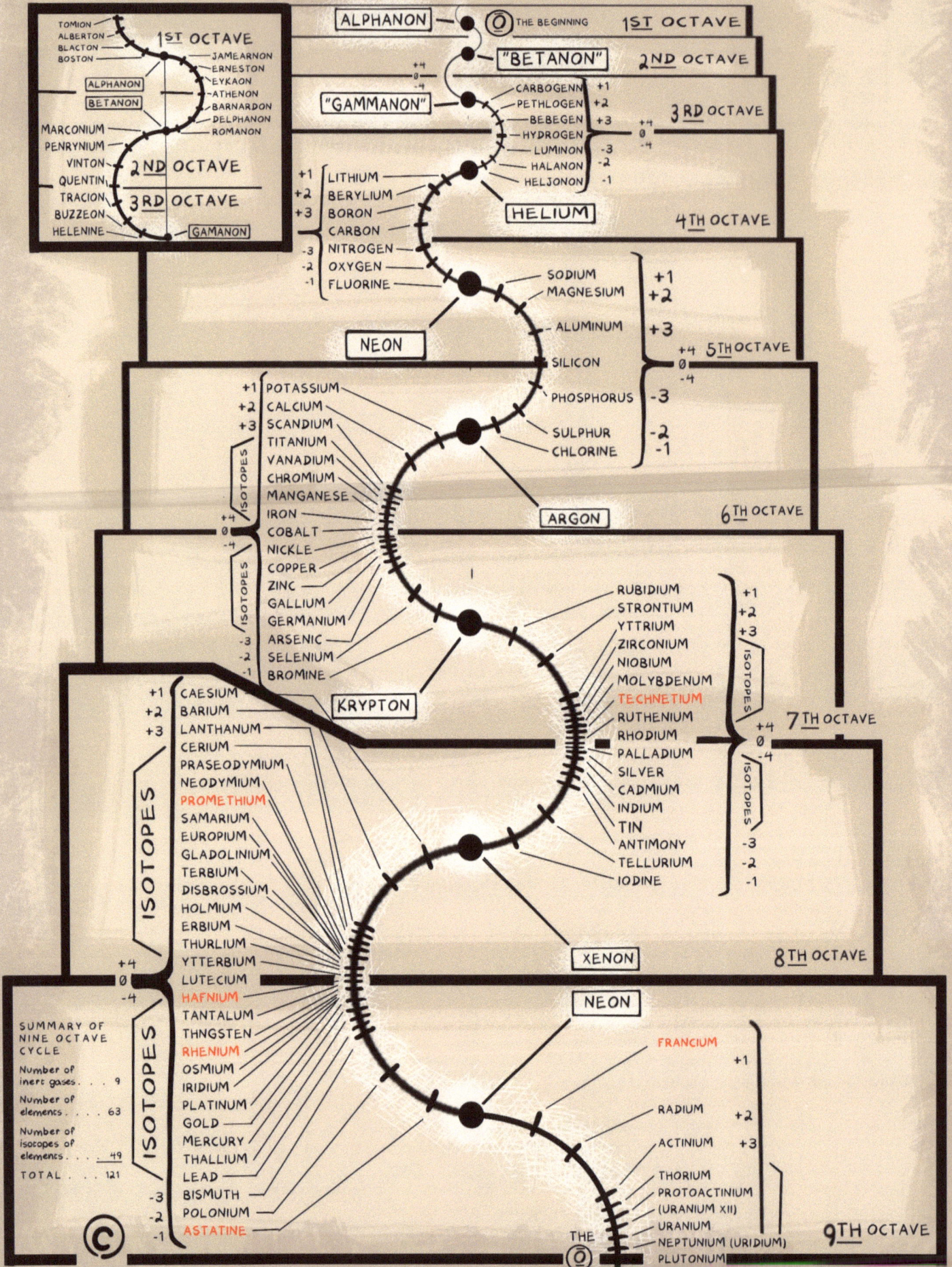

Inset box (top left):

1ST OCTAVE

TOMION
ALBERTON
BLACTON
BOSTON
JAMEARNON
ERNESTON
EYKAON
ATHENON
BARNARDON
DELPHANON
ROMANON

ALPHANON
BETANON

MARCONIUM
PENRYNIUM
VINTON
QUENTIN
TRACION
BUZZEON
HELENINE

2ND OCTAVE

3RD OCTAVE

GAMANON

Main diagram:

ALPHANON — ⊙ THE BEGINNING — 1ST OCTAVE

"BETANON" — 2ND OCTAVE
+4 / 0 / -4

"GAMMANON" — 3RD OCTAVE

CARBOGENN +1
PETHLOGEN +2
BEBEGEN +3
HYDROGEN +4 / 0 / -4
LUMINON -3
HALANON -2
HELJONON -1

+1 LITHIUM
+2 BERYLIUM
+3 BORON
CARBON
-3 NITROGEN
-2 OXYGEN
-1 FLUORINE

HELIUM

4TH OCTAVE

SODIUM +1
MAGNESIUM +2
ALUMINUM +3
SILICON +4 / 0 / -4
PHOSPHORUS -3
SULPHUR -2
CHLORINE -1

NEON

5TH OCTAVE

+1 POTASSIUM
+2 CALCIUM
+3 SCANDIUM
TITANIUM
VANADIUM
CHROMIUM
MANGANESE
ISOTOPES
+4 / 0 / -4 IRON
COBALT
NICKLE
COPPER
ISOTOPES
ZINC
GALLIUM
GERMANIUM
-3 ARSENIC
-2 SELENIUM
-1 BROMINE

ARGON

6TH OCTAVE

RUBIDIUM +1
STRONTIUM +2
YTTRIUM +3
ZIRCONIUM
NIOBIUM
MOLYBDENUM
ISOTOPES
TECHNETIUM
RUTHENIUM
RHODIUM +4 / 0 / -4
PALLADIUM
SILVER
CADMIUM ISOTOPES
INDIUM
TIN
ANTIMONY -3
TELLURIUM -2
IODINE -1

KRYPTON

7TH OCTAVE

+1 CAESIUM
+2 BARIUM
+3 LANTHANUM
CERIUM
PRASEODYMIUM
NEODYMIUM
PROMETHIUM
SAMARIUM
EUROPIUM
GLADOLINIUM
TERBIUM
DISBROSSIUM
HOLMIUM
ERBIUM
THURLIUM
YTTERBIUM
+4 / 0 / -4 LUTECIUM
HAFNIUM
TANTALUM
THNGSTEN
RHENIUM
OSMIUM
IRIDIUM
PLATINUM
GOLD
MERCURY
THALLIUM
LEAD
-3 BISMUTH
-2 POLONIUM
-1 ASTATINE

ISOTOPES

ISOTOPES

XENON

8TH OCTAVE

NEON

FRANCIUM +1
RADIUM +2
ACTINIUM +3
THORIUM
PROTOACTINIUM
(URANIUM XII)
URANIUM
NEPTUNIUM (URIDIUM)
PLUTONIUM

THE ⊙ END AND THE BEGINNING

9TH OCTAVE

Summary box (bottom left):

SUMMARY OF NINE OCTAVE CYCLE

Number of inert gases 9
Number of elements 63
Number of isotopes of elements . . . 49
TOTAL . . . 121

©

WALTER RUSSEL
1926

◄···· WALTER RUSSELL'S
PERIODIC CHART OF ELEMENTS

Walter Russell's Periodic Table differs from the traditional table we learned in school by integrating the concept of octave waves, reflecting nature's harmonic principles. In music, an octave is a doubling of frequency, and a perfect fifth is a ratio of 3:2. Walter proposed that elements and their isotopes follow a musical pattern similar to octaves and perfect fifths, arranging elements so that their atomic structures mirror these harmonic intervals. This approach suggests that elements transition smoothly through harmonic stages, reflecting an inherent order and symmetry in the universe.

Octaves and Elements:

• The table is divided into 9 "octaves," each representing a different stage in the progression of elements. This reflects Walter's idea that elements evolve through a musical scale-like pattern, much like the octaves in music.

• Each octave contains elements grouped together based on their properties and atomic structures as interpreted by Walter.

Vertical Progression:

• Elements are arranged vertically within each octave, moving from simpler to more complex structures as you ascend. For example, lighter elements like hydrogen and helium are in the first and second octaves, while heavier elements like gold and uranium are in the higher octaves.

Spirals and Connections:

• The central spiral line connects the elements, symbolizing the continuous and cyclical nature of elemental evolution. Each turn of the spiral represents the completion of an octave and the beginning of the next.

• The spiral suggests a dynamic and interconnected view of the elements, differing from the traditional static rows and columns of the standard periodic table.

Labeling and Symbols:

• Elements are labeled with their names and symbols, similar to the traditional periodic table. The elements are further annotated with their respective positions in the octaves.

• Isotopes and their relationships are also indicated, showing the variations of elements within the same octave.

144,000 /6.666 =21,600 [MOON D]
144,000 /5.555 =25,920 [GREAT YEAR]
144,000 /4.444 =32,400 [SUN R]
144,000 /3.333 =43,200 [SUN R]
144,000 /259.2 =555.555 [SD]
144,000 /16 =9,000 [BASE 9]
144,000 /12 =12,000 [BASE 24]
144,000 /266.666 =540 [POS]
144,000 /7.5 =19,200 [BASE 24]
144,000 /360 =400 [BASE AC/DC]
144,000 /2,160 =66.666 [ENERGY]

Aether is direct
current (DC)
Energy at rest

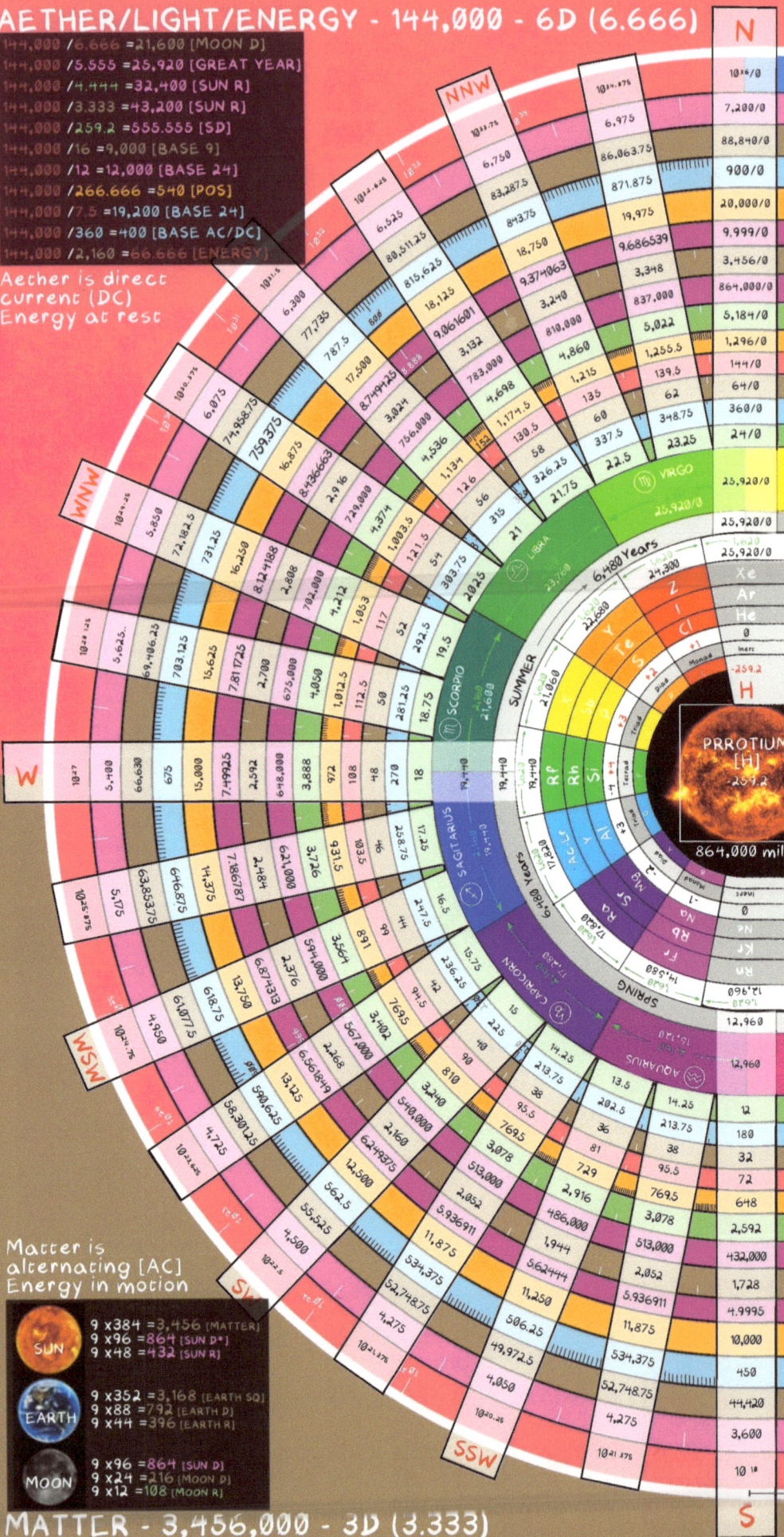

MALCOLM BENDALL'S
PLASMOID UNIFICATION MODEL

The Plasmoid Unification Model (PUM) integrates atomic, cosmic, and elemental concepts into a single framework, presenting a holistic view of matter and energy.

How To Read The PUM:

1. Plasmoids: Central to the model, these toroidal structures store and release energy, functioning as atomic batteries.

2. Elements: Listed with their respective frequencies and valencies, highlighting

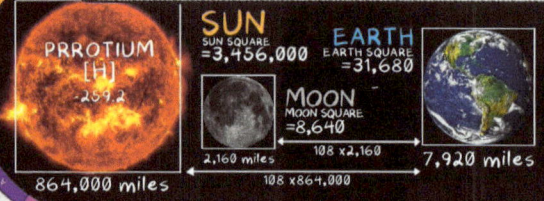

relationships based on harmonic resonance.

3. Cosmic and Elemental Scales: Connects solar, lunar, and atomic dimensions, showing the interrelationship between time, matter, and light.

4. Mathematical Constants: Key constants and their roles in calculating frequencies and dimensions.

5. Dimensions: From 3D matter to 6D light, each dimension has a specific frequency and role in the model.

PRROTIUM [H] -259.2

SUN
SUN SQUARE =3,456,000
864,000 miles

EARTH
EARTH SQUARE =31,680
7,920 miles

MOON SQUARE =8,640
2,160 miles   108 x 2,160
108 x 864,000

Matter is alternating [AC] Energy in motion

SUN
9 x384 =3,456 [MATTER]
9 x96 =864 [SUN D*]
9 x48 =432 [SUN R]

EARTH
9 x352 =3,168 [EARTH SQ]
9 x88 =792 [EARTH D]
9 x44 =396 [EARTH R]

MOON
9 x96 =864 [SUN D]
9 x24 =216 [MOON D]
9 x12 =108 [MOON R]

OUR CURRENT POSITION ON THE GREAT YEAR (25,920 Years) THE DAWNING OF THE AGE OF AQUARIUS

Comparison:

Mendeleev's Periodic Table:
- Organizes elements by atomic number and properties.
- Focuses on chemical behaviors and relationships.

Russell's Periodic Chart:
- Emphasizes a spiral structure, integrating natural cycles and cosmic principles.
- Highlights the energetic and vibrational aspects of elements.

Bendall's Plasmoid Unification Model:
- Combines atomic, cosmic, and vibrational principles.
- Integrates elements, cosmic cycles, and dimensions into a unified harmonic resonance framework.
- Adds plasmoids as a fundamental energy storage and conversion concept.

In summary, while Mendeleev's and Russell's tables focus on atomic properties and cosmic cycles, respectively, the PUM provides a broader, integrated perspective connecting atomic, cosmic, and energetic principles.

NOTE: The model is originally displayed as one circle. We split the model into two halves for a better display in the book. You can view the original by scanning this QR code:

www.strikefoundation.earth/open-source-research

LEGEND

Aether/Light · Frequency · Sun · Matter · Time · Degrees

- **Sun** (864,000) [DC], **Earth** (7,920) [AC], **Moon** (2,160) [AC]
- **Music** [AC] [Do=C=24x11.111=266.666]
- **Elemental Crystal Forms** +/- Monad, Diad, Triad, Tetrad
- **Elemental Valencies** (0,-1,-2,-3,-4 and 0,+1,+2,+3,+4)
- **Elements 1-16** (He-Cl)
- **Elements 17-32** (Ar-I)
- **Elements 33-48** (Xe-'Z')
- **Elemental Frequencies** (1,620x16=25,920 light frequency of -259.2 C)
- **Seasons of Great Year** (25,920/8 years)
- **Zodiac Great Year** (25,920/8 years)
- **Clock 24 Hour/0 Hour** [AC]/0 Hour [DC] Clock
- **Compass 360° Degrees** [AC]/0° Degrees[DC]
- **Matter** 64/0 64 Points being 32 Resonant Planes/0
- **Light** 144/0
- **Resonant Frequency Energy Unit (RFEU)** - 1,296(129,600)
- **All Time** 5,184(518,400 secs)
- **Aether-Sun** (864,000 miles diameter, 432,000 miles radius)
- **Matter** 3,456(3,456,000 miles-Sun Square)
- **Dimensions** 3D Matter=3.33, 4D Time=4.44, 5D Aether=5.55, 6D Light=6.666
- **Sound and Music** (0-20,000Hz)
- **Language** 1-9,10-90,100-900 [111,222,333,444,555,666,777,888,999][45,450,4,500]
- **Solar System** Sun & Planetary Diameters and Radii in miles
- **Plasmoids** 7,200 Degrees/0, 32 Planes, 64 Radial Points, One Zero Point
- **All Plasmoid Energy** =All Alternating Current [AC] Frequencies

(= Non Ionizing = Ionizing)

864,000 MILES IS THE SUN'S DIAMETER
86,400 SECONDS IN 1 EARTH DAY
8,640 MOON SQUARE, [4x2,160]
864,000 D x4 =SUN SQ. OF 3,456,000
3,456,000/16 =216,000[Moon D]
3,456,000/259.2 H =13,333.3[16/12]
3,456,000/6.666 =518,400[TIME]
TOTAL 864=846=684=648=486=468
864,000/25,920 =33.333[3D]
864,000/518,400 =1.666[POS]
864,000/129,600[RFEU] =6.666

Aether [DC] to Matter [AC] converter

Aether [DC] to Matter [AC] converter

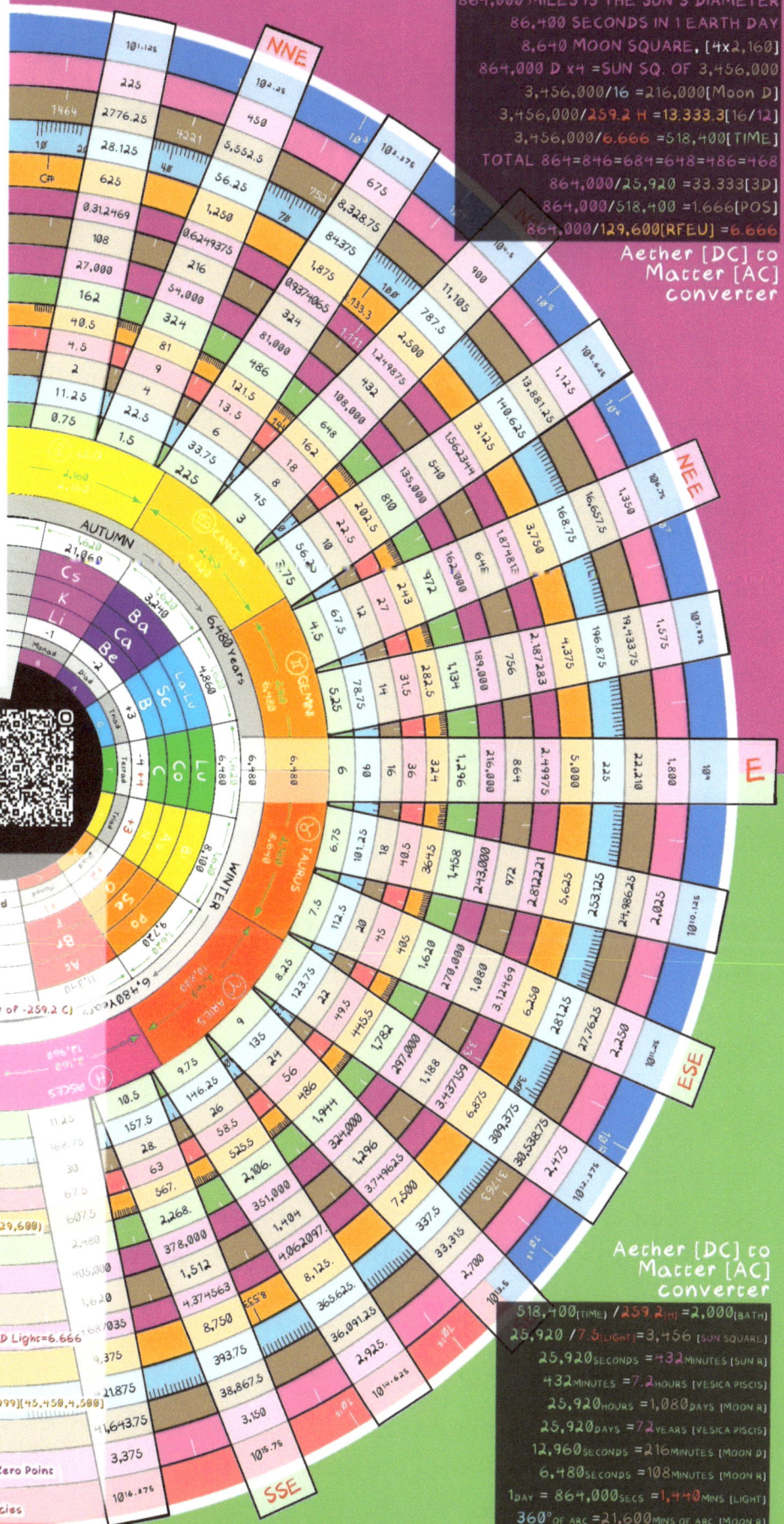

518,400[TIME] /259.2[H] =2,000[BATH]
25,920 /7.5[LIGHT] =3,456 [SUN SQUARE]
25,920 SECONDS =432 MINUTES [SUN R]
432 MINUTES =7.2 HOURS [VESICA PISCIS]
25,920 HOURS =1,080 DAYS [MOON R]
25,920 DAYS =72 YEARS [VESICA PISCIS]
12,960 SECONDS =216 MINUTES [MOON D]
6,480 SECONDS =108 MINUTES [MOON D]
1 DAY = 864,000 SECS =1,440 MINS [LIGHT]
360° OF ARC =21,600 MINS OF ARC [MOON R]
21,600 MINS OF ARC =1,296,000 [SECS RFEU]

GLOSSARY

DEFINITIONS AND EXPLANATIONS

304 Alloy Steel: Referred to as 304 stainless steel, 304 alloy is a widely used austenitic (non-magnetic) stainless steel.
(page 18)

4:3:2: The ratio 4:3:2, in relation to sacred geometry, harmonies, and musical intervals like the perfect fourth and perfect fifth, embodies significant mathematical and philosophical concepts that have fascinated scholars, musicians, and mystics for millennia. Specifically in music, the perfect fourth and perfect fifth are intervals that have been considered harmonious since ancient Greece and serve as fundamental building blocks in the construction of musical scales. The mathematical ratios that define these intervals are:

• <u>Perfect Fifth</u>: The ratio of frequencies between the notes of a perfect fifth is 3:2. For example, if one note vibrates at 300 Hz, the note a perfect fifth below it vibrates at 200 Hz.
• <u>Perfect Fourth</u>: The ratio for a perfect fourth is 4:3. If one note vibrates at 300 Hz, the note a perfect fourth above it vibrates at 400 Hz.
(pages 17, 27, 34)

Atomic Orbitals: In chemistry and physics, atomic orbitals describe the regions around an atom's nucleus where electrons are most likely to be found. Think of them as maps that predict where an electron is likely to be located at any given time, based on quantum mechanical calculations. Each orbital can hold a maximum of two electrons and has a unique shape and energy level, which influence how atoms interact with each other.
(page 45)

Atoms: Atoms are the smallest units of an element and are considered its building blocks. Each atom consists of a nucleus, made up of protons and neutrons, surrounded by a cloud of electrons. The protons carry a positive charge, electrons carry a negative charge,

["Atoms" continued]

and neutrons have no electric charge. The interactions between these charged particles give substances their chemical properties.
(pages 17, 23)

Carbon Monoxide: Abbreviated as CO, carbon monoxide is a colorless, odorless, and tasteless gas that can be deadly. It's the leading cause of poisoning deaths in the United States, killing hundreds of people each year and making thousands more sick. CO is harmful because it displaces oxygen in the blood, depriving the heart, brain, and other vital organs of oxygen. This can lead to loss of consciousness and suffocation within minutes. A lethal dose is greater than 400 ppm. If in an enclosed room, evacuate immediately.

SYMPTOMS: headache, fatigue, dizziness, drowsiness, nausea, chest pain, vomiting, confusion, muscle weakness, and collapse.
(pages 16, 20)

Carburetor: A carburetor is a mechanical device in an internal combustion engine that mixes air and fuel in the correct ratio for combustion. It's an essential component in older vehicles and machinery that don't use modern fuel injection systems.
(pages 27, 31, 34)

Frequency: Frequency refers to how many times something completes a vibration cycle (or oscillation) in one second. This is called the "frequency" of the vibration. For instance, if you pluck a guitar string and it moves back and forth 432 times in one second, it's vibrating at a frequency of 432 cycles per second. We measure frequency in units called hertz (Hz), where 1 Hz represents one complete vibration cycle per second.

Understanding frequency is key in various fields, including music and physics. In music, the frequency of the sound produced by an instrument determines the pitch we hear. Think about the difference in sound between a flute (higher frequency) and a tuba (lower frequency). In physics, frequency helps us analyze patterns in sound, light, and radio waves. Higher frequencies mean more cycles occur in one second, leading to different properties and effects, such as a higher-

pitched sound or a different color of light.
(pages 16, 17, 27, 43)

Harmonics: In physics and music, harmonics are frequencies that are integer multiples of a fundamental frequency, which together contribute to the overall sound of a musical note or any vibrating system. The fundamental frequency is the lowest (first) harmonic and determines the pitch of the sound, while the higher harmonics (second, third, etc.) add texture and complexity, influencing the timbre or color of the sound. For instance, if the fundamental frequency of a note is 100 Hz, its second harmonic would be 200 Hz, the third harmonic 300 Hz, and so on.
(pages 17, 27)

Imperial Measurement System: The imperial measurement system of inches, feet, yards, and miles, with its divisions by 12 (inches in a foot), more directly aligns with the numerical values of sacred geometry than the metric system, which is based on units of 10.
(page 17)

Ionization: This occurs when ultraviolet (UV) light gives atoms a powerful energy boost, exciting electrons to break free from their orbits. When atoms lose electrons and therefore gain a positive charge, they become ions. Our sun is doing this naturally all around us.
(pages 15, 23, 25, 27, 32)

Isotope: An isotope is a variant of a chemical element that has the same number of protons (which defines the element) but a different number of neutrons in its nucleus. This means that while isotopes of an element share the same atomic number, they have different atomic masses. Physical properties such as density, melting point, and boiling point can vary between isotopes due to differences in mass.
(page 43)

Micron: A micron, or micrometer, is one millionth of a meter. For reference, a fine human hair is around 50 microns wide.
(pages 17, 25)

Noble Gases: Noble gases are a common group of elements in the periodic table that rarely interact with other elements. This is because their outer electron shells are full, so they don't need to form bonds with other atoms to become stable. These gases include helium, neon, argon, krypton, xenon, and radon.
(page 41, 45)

Plasma: Plasma is often described as the fourth state of matter, distinct from the more familiar solid, liquid, and gas states. It's a hot, ionized gas consisting of an approximately equal number of positively charged ions and free electrons, which gives it unique properties and behaviors. Unlike gases, where atoms are whole and electrically neutral, plasma contains atoms that have been energized to such an extent that their electrons are freed from their orbitals. The creation of plasma typically involves adding energy to a gas, such as by heating it or subjecting it to a strong electromagnetic field, to the point where the electrons escape from the atoms. This ionization process results in a collection of charged particles that can conduct electricity and respond dynamically to electromagnetic fields.
(pages 27, 32)

Plasmoids: Plasmoids are essentially self-contained clusters of plasma that have magnetic field lines wrapped around them, which can be ejected from or move within larger plasma environments.
(pages 15, 25, 31)

Pistons: Pistons are crucial components of the internal combustion engines found in vehicles and other machinery. They move up and down inside the engine's cylinders, creating a reciprocating motion that is converted into a rotational motion by the crankshaft. This process begins when the engine's fuel-air mix, introduced by the carburetor, is ignited, causing an expansion of gases that pushes the piston down. The piston itself is connected to the crankshaft through a connecting rod, and as the piston moves, it turns the crankshaft, which ultimately drives the vehicle's wheels.
(pages 15, 16, 23)

Ratio: A mathematical ratio represents a relationship between two or numbers that shows how many times one value contains or is contained

["Ratio" continued]

within the other. It is expressed as "a to b" or a:b, with "a" and "b" representing the quantities being compared. Ratios can be used to compare quantities of the same kind.

For example, if there are 2 apples and 3 oranges, the ratio of apples to oranges is written as 2:3, indicating that for every 2 apples, there are 3 oranges.

(page 27)

Retrofitted: To retrofit means to add new technology or features to older systems, buildings, machinery, or equipment, often for the purpose of making them more efficient, effective, compliant with current standards, or capable of new functionalities.

(pages 15, 20, 31)

Sacred Geometry: Sacred geometry serves as a cross-disciplinary bridge connecting the realms of mathematics, art, philosophy, and spirituality, suggesting that universal patterns and ratios are inherent in the very fabric of the natural world and are reflective of broader cosmic truths.

(page 27)

Sanskrit Mathematics: These are mathematical concepts and techniques that originated in ancient India, documented in texts like the Sulbasutras (800-500 BCE), Aryabhatiya by Aryabhata (499 CE), and Brahmasphutasiddhanta by Brahmagupta (628 CE). These works cover a range of topics, including arithmetic, algebra, geometry, and trigonometry, and introduced significant concepts such as zero, the decimal system, and detailed solutions to quadratic equations. The decline of Sanskrit mathematics was due to several factors, including the influence of British colonization, which introduced Western mathematical frameworks and education systems, and the global standardization of mathematics, which favored Western methods. Despite this decline, the influence of Sanskrit mathematics persists in some modern principles and continues to be of historical and academic interest.

(Page 17)

Ultraviolet: UV light refers to electromagnetic radiation with wavelengths shorter than visible light but longer than X-rays, specifically in the range between 10 nanometers (nm) to 400 nm. When we talk about UV light in reference to its "vibration" at a specific wavelength, like 100 nm, we're discussing its frequency or the energy associated with that wavelength. At 100 nm, UV light is well within the range classified as "far UV" or "vacuum UV," where air readily absorbs it. This part of the UV spectrum has higher energy compared to UV light with wavelengths closer to the visible spectrum (near UV, around 400 nm). Given its shorter wavelength (100 nm), the UV radiation at this point has higher frequency and thus higher photon energy, making it capable of breaking chemical bonds and ionizing atoms and molecules more efficiently than UV light at longer wavelengths.
(pages 15, 23)

Vajra: The name references a significant symbol in Hinduism and Buddhism, often associated with both spiritual and physical power. In Sanskrit, "Vajra" means both "thunderbolt" and "diamond."

Vortex: A vortex is a flow phenomenon characterized by the circular or spiral motion of a liquid, gas, or plasma around a central axis. The rotational movement in a vortex causes the velocity to be fastest near the core and to decrease with distance from the center. The pressure is typically lowest at the core for free vortices (not driven by external forces), leading to phenomena like the eye of a hurricane where calm conditions can be found at the center of the storm.
(page 27)

BIBLIOGRAPHY

WITH QR CODES

ALCHEMICAL SCIENCE
EST. April 12, 2023 on YouTube (https://youtube.com/@alchemicalscience)

01 "Thunderstorm Generator | COMPLETE DIY BUILD GUIDE | Malcolm Bendall's Plasmoid Tech" by Alchemical Science, https://youtu.be/l6cxYnWTFi0 [November 5, 2023]

02 "Thunderstorm Generator Build Guide Updates from MFMP | Plasmoid Tec" by Alchemical Science, https://youtu.be/UbenNUFsWYA [November 13, 2023]

03 "Plasmoids / Ball Lightning | NEW EARTH SHATTERING FINDINGS + The Full Story so far..." by Alchemical Science, https://youtu.be/6VFJuIpHckY [April 3, 2024]

04 "Malcolm Bendall's Thunderstorm Generator | BEGINNERS GUIDE + NEW Trial Footage from the US" by Alchemical Science, https://youtu.be/yN5fz8QgDuY [April 25, 2024]

05 "Introduction to Plasmoids, Ball Lightning & Thunderstorms" by Alchemical Science, https://youtu.be/DrlL2N1A4GE [January 28, 2024]

06 "Malcolm's Thunderstorm Plasmoid Generator in Action | FULL DOCUMENTARY | with Jordan & Roland Perry" by Alchemical Science, https://youtu.be/-ugB_nK-Mu0 [September 12, 2023]

07 "Thunderstorm Generator Q & A | More Proof of Cavitation from MFMP | Plasmoid Tech Updates" by Alchemical Science, https://youtu.be/E-z12wCWG1c [March 16, 2024]

08 "Thunderstorm Generator Live Demo and Trials at Cosmic Summit 2024 | Much more to come" by Alchemical Science, https://youtu.be/DQ6D2XBJbXM [June 29, 2024]

MARTIN FLEISCHMANN MEMORIAL PROJECT

EST. 2012 on https://www.quantumheat.org/

09 "THOR", written by Bob Greenyer, http://www.quantumheat.org/index.php/en/home/mfmp-blog/560-thor [October 3, 2023]

10 "THOR - Live Thunderstorm generator demonstration under load in Technopark, Zurich conference hall" by MFMP, https://youtu.be/m-h4N94V5s0 [October 15, 2023]

STRIKE FOUNDATION

EST. 2022 on https://www.strikefoundation.earth/

11 "OPEN SOURCING THE NOTES OF INVENTOR MALCOLM BENDALL" by Samadhi Halkett Lewis, https://www.strikefoundation.earth/open-source-research [March 2024]

KONCRETE PODCAST

EST. July 2014 on YouTube (https://youtube.com/@koncrete)

12 "Randall Carlson Finally Reveals Proof of Ancient Lightning Bolt Technology" by Koncrete Podcast, https://youtu.be/1ogPBJyDFWw [Sep 25, 2023]

AFTER SKOOL

EST. July 2016 on YouTube (https://youtube.com/@AfterSkool)

13 "HIDDEN MATHEMATICS - Randall Carlson - Ancient Knowledge of Space, Time & Cosmic Cycles" by After Skool, https://youtu.be/R7oyZGW99os [Dec 8, 2020]

"The day science begins to study non-physical phenomena, it will make more progress in one decade than in all the previous centuries of its existence."

Nikola Tesla

www.ingramcontent.com/pod-product-compliance
Lightning Source LLC
Chambersburg PA
CBHW060806270326
41927CB00002B/68